CENTRIFUGAL DRYERS AND SEPARATORS

DESIGN & CALCULATIONS

BY

Eustace A. Alliott

Wexford College Press

2006

CONTENTS

LIST OF ILLUSTRATIONS

1*   ix

LIST OF ILLUSTRATIONS

NOTE ON UNITS

THE foot pound second system is used throughout this book, except for various equations in Chapter IV, where centimetre gram second units are more convenient.

All forces expressed in the familiar gravitational units (pounds or grams) are reduced to fundamental units by multiplying by g./32·2 ft. per sec.², or 981 cms. per sec.². The basic equation for force is therefore :—

$$\text{Force} = \text{acceleration} \times \text{mass}.$$

c.c. $$\frac{\text{Pounds} \times \text{g.}}{\text{(Poundals)}} = \text{ft. per sec.}^2 \times \text{lb.}$$

or $$\frac{\text{Grams} \times \text{g.}}{\text{(Dynes)}} = \text{cms. per sec.}^2 \times \text{grams.}$$

CENTRIFUGAL DRYERS AND SEPARATORS

CHAPTER I

CENTRIFUGAL FORCE AND BASKET STRENGTH

Centrifugal force—Surface speed—Pressure of solid and liquid loads—Self stress—Total stress—Speeds and effects.

THE whole operation of a centrifugal dryer or separator centres round the rotating drum or basket. This is perforated in the case of dryers, but separator drums are plain.

In each case the object is to intensify the effects normally produced by gravity, for which a much greater centrifugal force is substituted. It is important, therefore, to know the effect produced, and to understand its relation to the speed of the basket, and the stresses produced on it. This effect is most conveniently measured by the ratio of the centrifugal force to gravity, or by the force in pounds exerted by a mass of one pound on the basket periphery.

If we write C_{POUNDS} for this force we have according to the ordinary rule

$$C_P g = \tfrac{1}{2}\omega^2 D_P . \qquad . \qquad . \qquad . \quad (1)$$

Whence $\qquad C_P = \dfrac{D_P N^2}{5866} \qquad . \qquad . \qquad . \quad (2)$

Here D_P = dia. of basket periphery in feet.

$g = 32 \cdot 2$.

N = Revs. per min.

ω = radians per sec.

11

There is an interesting relationship between the centrifugal effect and the linear speed, V_P, of the periphery.

$$V_P = \pi D_P N \quad . \quad \quad . \quad \quad . \quad \quad . \quad (3)$$

or $\quad C_P = \dfrac{V_P^2}{57888 D_P} \quad . \quad \quad . \quad \quad . \quad (4)$

Figs. 1 and 2 illustrate the relationship between C_P and V_P for various sizes and speeds of basket.

The basket may be loaded either with solid or liquid material.

For a true solid load, in which one would suppose the grains to have an angle of repose of 90°, the pressure exerted is the total " centrifugal effort " divided by the total area of the basket rim. In the case of a liquid load, the " effort " of each elementary ring is transmitted to the rim as a hydrostatic pressure, so that the smaller " effort " of portions nearer the centre of the basket is compensated by the increased ratio between the diameter at which the initial pressure is exerted and the diameter of the basket wall.

If C is the centrifugal force at any diameter D feet, then we have

$$C = \dfrac{C_P}{D_P} \times D . \quad \quad . \quad \quad . \quad \quad . \quad (5)$$

If we assume that the basket is not filled to the centre, and the load has an inner diameter D_L, weighs S_L lb. per cubic foot and is divided into a very large number of concentric rings of very small thickness $\frac{1}{2}dD$; we then have the following

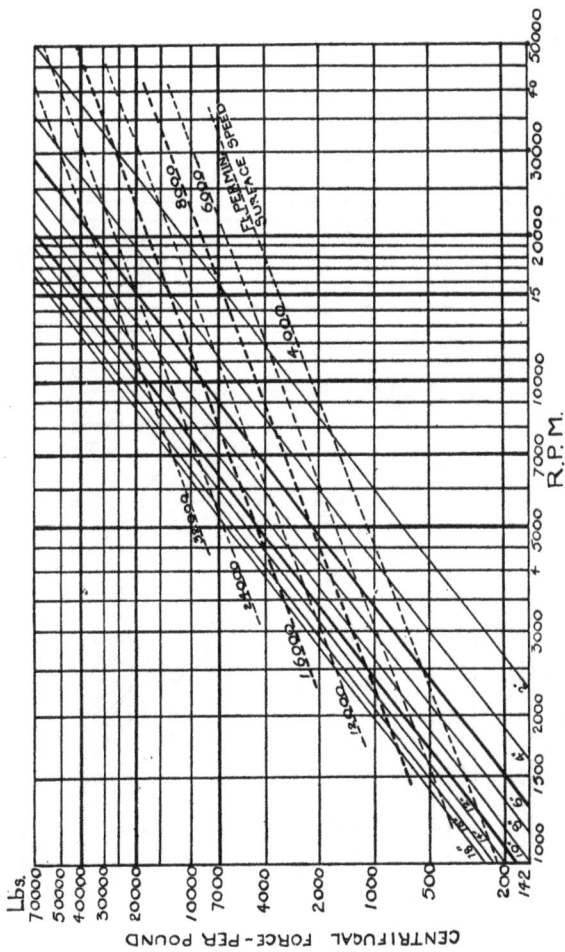

Fig. 1.—Relation between Centrifugal Force, Surface Speed and Diameter. Sizes 2″—18″.

13

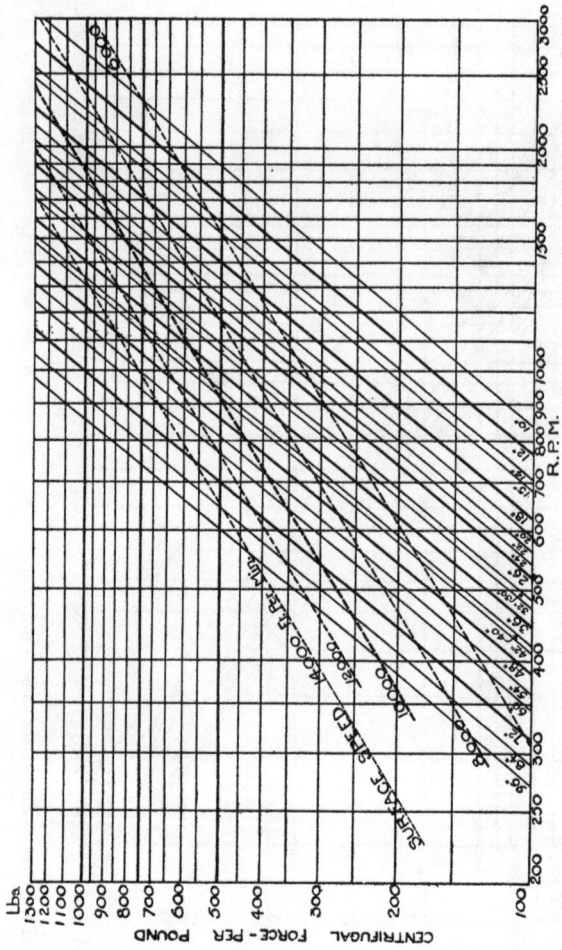

Fig. 2.—Relation between Centrifugal Force, Surface Speed and Diameter. Sizes 10″—96″.

expression for the mass dM_1 of each elementary ring, taking a 1 ft. section of the basket :—

$$dM_1 = \tfrac{1}{2}(\pi D S_L) dD.$$

The total weight of load, per foot of basket depth, is :

$$M_1 = \tfrac{1}{4}\pi S_L(D_P{}^2 - D_L{}^2) \quad . \quad . \quad . \quad (6)$$

The pressure P_S exerted by a solid load on the rim, in pounds per sq. in., is, of course, the total " centrifugal effort " of all the elementary rings, divided by the peripheral area in square inches. For one ring, therefore, we have :

$$dP_S = \frac{C_P D}{D_P} \times \tfrac{1}{2}(\pi D S_L) dD \div 144\pi D_P$$

$$P_S = \frac{C_P S_L}{288 D_P{}^2} \int_{D_L}^{D_P} D^2 dD$$

or $\qquad P_S = \dfrac{C_P S_L(D_P{}^3 - D_L{}^3)}{864 D_P{}^2} \qquad . \quad . \quad (7)$

In the case of a liquid load the elementary pressure produced by each ring is its " centrifugal effort " divided by the surface of the element.

Thus $\qquad dP_L = \dfrac{C_P D}{D_P} \times \tfrac{1}{2}(\pi D S_L) dD \div 144\pi D$

or $\qquad P_L = \dfrac{C_P S_L}{288 D_P} \displaystyle\int_{D_L}^{D_P} D\, dD$

Whence $\quad P_L = \dfrac{C_P S_L(D_P{}^2 - D_L{}^2)}{576 D_P} \qquad . \quad . \quad (8)$

Now from (6) we see that

$$S_L(D_P{}^2 - D_L{}^2) = \frac{4 M_1}{\pi}$$

whence $\qquad P_L = \dfrac{C_P M_1}{144\pi D_P} \qquad . \quad . \quad . \quad (9)$

The latter equation, of course, is the value obtained by a plain calculation assuming the whole mass of the liquid to be concentrated at the rim.

It is interesting to note that in a 36-in. machine of the ordinary type, running at 1000 r.p.m. and producing an effect of about 511, the pressure produced by a 4-in. wall of water is about 62 lb. per sq. in. This would rise to about 166 lb. per sq. in. if the machine were filled entirely up to the central axis.

In a very special machine only 2 in. in diameter running at about 37,500 r.p.m., and producing an effect of 40,000, the pressure for a $\frac{1}{2}$-in. wall of water would be about 540 lb. per sq. in.; or 720 lb. per sq. in. if entirely full.

In actual practice, loads of the ideal solid type are extremely rarely met with. When dealing with loose granular substances, or muds and so forth, it is well to assume, in most cases, for the sake of safety, that pressure follows the rule for liquids. In a given substance it is, of course, purely a question of the angle of repose. In textile and similar materials, the load is to a certain extent self-supporting and takes up a good deal of the strain within its own structure. This is evidenced by baskets which come back to the makers to be overhauled after very long periods of use and inattention. Many of these are found on examination to be in such a state that had the periphery been subjected to the whole calculated stress it must certainly have given way.

The calculation of the strength of the centrifugal basket has certain analogies with that for a boiler. There is, however, an important additional factor in the case of the centrifugal, which renders it uneconomical, as a rule, to attain greater pressures, involving higher speeds, by merely increasing the thickness of the rim. The weight of the periphery itself must be taken into account and for an unloaded basket of a given material there is, as will be seen below, a certain surface speed at which bursting will occur, which is independent of the thickness of the rim or its diameter. There is thus a close analogy with the case of a belt pulley or a fly-wheel.

Let S_P be the *average* weight of the periphery, in lb. per cu. in., after making any necessary deduction for perforations. Let H_P and H_D be the pitch and diameter of the holes, in inches, T_P the tensile stress in lb. per sq. in., Z the thickness of the periphery in inches, P_P the pressure per sq. in. on the periphery due to its own mass.

$$P_P = C_P S_P Z$$

$$\therefore \quad T_P = \frac{H_P}{H_P - H_D} \times \frac{12 D_P C_P S_P Z}{2Z}$$

$$T_P = \frac{6 H_P D_P C_P S_P}{H_P - H_D} \quad . \quad . \quad . \quad (10)$$

or, from (4)

$$T_P = \frac{H_P}{H_P - H_D} \times \frac{V_P^2 S_P}{9648} \quad . \quad . \quad (11)$$

This confirms that the self-stress is solely dependent on the peripheral speed, and Fig. 3 shows graphically the amount of self-stress for various cases.

The total stress is, of course, the sum of the self-stress and that due to the pressure of the

material given by (7) or (9) and calculated by
the ordinary rule for boiler shells. The working
out of examples is best left to the reader, since the
process is simple, and the formulæ are cumber-
some. It suffices to say that in determining the
thickness of the basket wall an ample factor of

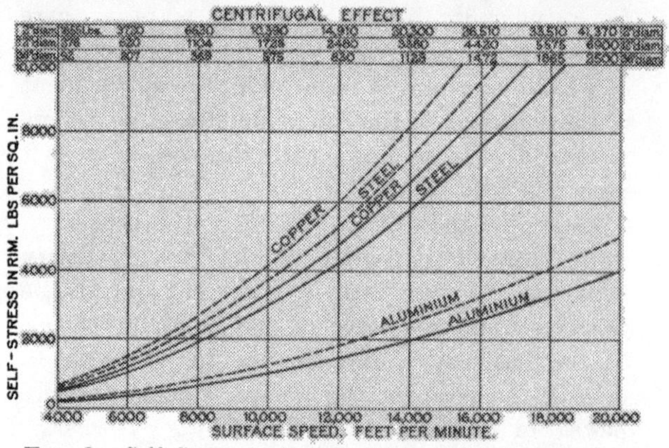

FIG. 3.—Self Stress in Basket Rim at Various Surface
Speeds.

No load. Compared with the centrifugal force in 2″,
12″ and 36″ diameter machines respectively. It is
assumed that the perforations increase the stress by 25%.
Full lines show stress in a rim which is not perforated.

safety should be used. This should be greater
in the case of a dryer than a separator, as the
loading in the former type is apt to be irregular,
setting up stresses which cannot be adequately
covered by any theoretical calculation. An allowance for corrosion should be added to the thickness
thus determined, according to the circumstances
of the case. The top and bottom plates of the
basket help to stiffen the periphery, and additional

assistance may be given by adding one or two
strengthening bands. In the case of very small
diameter baskets the tension is a combination of
that in a thick pipe with a somewhat analogous
one due to the self-stress.

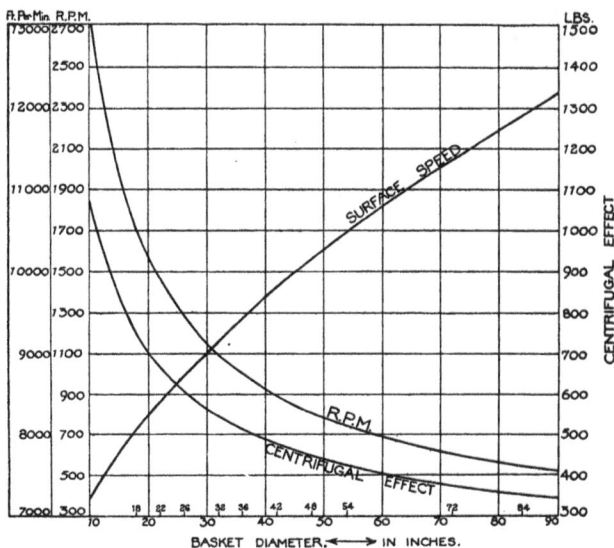

Fig. 4.—Chart showing Relations between Size, Speed and
Effect of Centrifugal Baskets.

As ordinarily based on a 36″ basket at 1000 r.p.m.

Investigation shows that if the load is a liquid
one, and is proportional to the diameter and
depth of the basket and the wall thickness, equal
stresses are given by equal peripheral speeds. If
a series of baskets was designed on this basis the
centrifugal effect would be inversely proportional
to the diameter, so that a 36-in. machine would

only give half the effect of an 18-in. machine, and a 72-in. only half that of a 36-in. machine. A rough balance is usually struck between the claims of equal peripheral speed and equal effect, and if the basis of design were, say, a 36-in. machine running at 1000 r.p.m., the speeds of other baskets would be much as shown in Fig. 4, which also indicates the relative effects. In practice, some other minor considerations would enter, and the curves would be more or less irregular, but the principle would still hold that the smaller machines would give higher effects at lower peripheral speeds.

CHAPTER II

Factors affecting time and dryness—Loading—Washing—
Experiments—Comparison with presses, etc.

IT is not easy to theorize on the length of time
required to dry any material in a centrifugal
machine. The operation is in many respects
similar to that of filtration, but it is possible to
handle substances of much coarser texture than
could be dealt with in any form of press.

The main factors are the following :—

(1) Particle size, or texture of the solid
matter.

(2) The viscosity of the liquid portion.

(3) The temperature at which the operation
is carried out.

(4) The initial proportion of liquid.

(5) The final dryness required.

(6) The centrifugal force employed.

(7) The thickness of layer.

We may consider, in the first instance, the case
of very coarse-grained materials including large
pea crystals. These do not ordinarily come to
the machine with any large proportion of water
and are often laid in the basket while the latter
is at rest. The pores are large and the bulk of
any liquid present immediately escapes on starting
up the machine. The exposed surface is small in
comparison to the volume and the points of
contact are comparatively few. There is therefore little opportunity for liquor to cling, and any
capillary or surface effects are small. Air can
pass freely through the mass and will have an

appreciable effect on the small amount of residual moisture. Under these conditions it is sufficient for most purposes merely to bring the basket up to full speed and then stop, for longer runs will only be necessary where requirements are somewhat stringent. With this class of material final moistures under 1% are quite feasible unless the liquor viscosity is high.

Materials of somewhat smaller grain size, say 1 to 3 mm. overall, are somewhat similar in their behaviour, but are more often fed as a slurry and may contain an appreciable amount of water. Examples are large crystals of sulphate of ammonia, crystal sugar, etc. One to 5 minutes may be allowed after full speed has been obtained and moistures from 1% to 2% are very usual. Another class of small grain particle is represented by vacuum salt, fine Epsom salts, fine sulphate of ammonia, etc. The size here may vary from 0·1 mm. up to 1 mm. Such products are nearly always fed as a slurry, and 50% of liquor in the entering material will be quite a usual figure. Capillary and surface film effects begin to be of more importance and air drying plays a minor, though still perceptible, part. Five to 10 minutes is a very usual drying time and final moistures would ordinarily vary from 2% to 5%. Between these and sludges, such as caustic mud, there are all kinds of gradations. In the latter substance we get quite a number of fine particles of not much greater diameter than μ, although a considerable number of larger agglomerations are present. The time required lengthens according to conditions up to, say, 20 minutes after the completion of acceleration or of the charging of the basket, and the final moisture may easily be anything up to 30% or 40%. Caustic mud, for instance, requires, say, 15 minutes' run and its moisture is often 30%, sometimes only 25%. In such a case the effect of air penetration is of no practical consequence and capillary and surface effects play a great part in retaining liquor.

It follows, therefore, that it is important that grain size and texture should be controlled as far as possible by the right selection of conditions for grinding or precipitation wherever such control is feasible. Something may be done by the use of right temperatures, concentrations and pressures in carrying out reactions, and prolonged digestion of precipitates at elevated temperatures in contact with the mother-liquor is usually of assistance in eliminating slimy tendencies.

If the particles are reasonably uniform in size, results will be appreciably better than if, say, a proportion of very fine particles were mingled with the coarser bulk.

Textiles form a class to themselves. The exposed surface is usually large in proportion to the weight of material, and the fabric is compressible, the interstices are small, and the fibres hollow. The moisture therefore is in general rather high, and is held in great part inside the fibres, and not on the surface. In laundry work it is usual to run sheets and unstarched work from 6 to 15 minutes and table linen and other starched work for a period of 12 to 25 minutes. Woollen goods should be spun for a very short time, and if they are in the machine for 2 or 3 minutes after the latter has obtained full speed it is generally sufficient, for unduly prolonged exposure to high centrifugal force will cause shrinkage. The time required varies so considerably from laundry to laundry, even though conditions are apparently identical, that personal idiosyncrasies on the part of the staff, or varying efficiencies in the subsequent drying operations, may not form the whole explanation. There appears to be a need for further research, since some laundries may run starched work as little as 4 minutes, and others are not content with less than the maximum time given above. It is possible that the nature of the washing operation, and in particular the presence or absence of lime soaps, may make an appreciable difference to the operation in the centrifugal. An average moisture in a London laundry would be about 32% to 33% as measured by the loss in weight after passing through the ironing machine. The problems presented by cotton yarn are not dissimilar, but the final moisture obtained is determined, not only by the class of yarn, but by such factors as the dye employed, etc.

Organic substances such as fish and meat offal, oil seeds, etc., are also handled in centrifugal dryers. In such cases the materials are usually cooked or predigested in order to rupture the cells and make it easier to extract any fat or oil. Steaming in the centrifugal is often resorted to in order to assist this action.

Yeast is sometimes passed through a centrifugal

in order to free it from beer, and will form a layer up to 5 in. in thickness, although possibly half this amount will be a fairer figure for an average run. Fifteen or 20 minutes' drying after filling will be usual, and the residual beer left in the yeast will be quite negligible. The filtered beer is apt to be strongly aerated in its passage through the machine, and emerges as a thick foam which has to be subsequently allowed to settle down in tanks. Still another class of problem is presented by metallic swarf and chips, which require to be freed from oil which otherwise would be wasted. Coarse chips yield less than the fine, and in one case coarse steel chips gave 1 gallon per cwt. while fine chips gave 4 gallons. The residual oil after spinning is so small as to be negligible. In this case, curiously enough, the fine swarf retains the least oil, since it packs closer than the coarse material, which still presents a large surface, much of which is adapted to catch the oil.

In most cases the presence of oil or any liquor of high viscosity slows the rate of flow very appreciably. Not only is more time required, but the residual liquor may be much greater. Increased temperature in such cases is often useful, since by its means the viscosity is reduced with beneficial effect. The use of steam for treating organic substances containing grease has been already mentioned. It is also employed for treating sugar and for dealing with oily rags and so forth. The latter may first be spun and steamed in order to free them from the bulk of the oil. Then they are washed in an ordinary washing machine and finally spun again to free them from water, and after this they are dried to make them ready for use. Steaming is also helpful where only water has to be removed, for even here a reduction of viscosity may be of some consequence. In steaming rags and so forth and other textile materials, it is possible to get over 5% reduction as compared with ordinary methods, but it is questionable whether the gain is worth the cost of the additional steam. It is not desirable in such cases to steam over the whole operation, as moisture condenses in the fabric and the real benefit is only obtained if the last five minutes or so of the operation are conducted without steaming. Hot water is useful in some operations where too high a temperature would be harmful, and helps in dealing with such substances as anthracene and naphthalene. Increased temperature helps air drying both in the basket if the material is open and subsequently if the latter is discharged on to an open conveyor. When the grains are of a fair size, the proportion of residual moisture removed by this means may be quite appreciable. It is quite feasible to supply hot air to the basket if the products are such as to permit free circulation and by this means a high degree of dryness can sometimes be obtained. The method is useless if there is much moisture which cannot be removed by centrifugal force alone, as in the case of fabric. Although it is generally believed in laundries that the circulation even of cold air must make an appreciable difference, it is extremely doubtful if this is so, for the amount passing through such goods is particularly small. The dirt which collects on a foggy day is often adduced as a proof of air penetration, but this is merely an instance of centrifugal settlement and would take place even though the basket walls were not perforated.

The proportion of liquid present in the entering material is only of serious importance in thin slurries of a slow filtering quality. For instance, 7 to 8 minutes may be required for filling a machine with caustic mud containing, say, 55% to 60% of liquid. In filtering yeast, 20 to 30 minutes may be required for filling, and this is true of picric acid and many other substances. On the other hand, if filtration proceeds freely and the proportion of liquid is not excessive, variations in its initial amount are almost of no consequence, since they will ordinarily be equalised almost as quickly as the product can be brought up to speed.

The effect of prolonged drying time on the final moisture follows a law of diminishing returns, particularly in the latest stages. In one instance, the increase in the time of treatment of cotton sheets from 10 to 20 minutes decreased the final moisture by some 2% to 3% only. Caustic mud had its final moisture reduced to from 34% to 26% by increasing the time of spinning from 8 to 16 minutes. The following results on vacuum salt are interesting. The time of treatment is reckoned after the completion of acceleration :—

Minutes.	% dryness.	Minutes.	% dryness.
0	5·14	7	2·25
$3\frac{1}{2}$	3·72	14	2·03

The question is often asked whether drying is not complete before the discharge spout has ceased to drip. This entirely depends on the degree of final moisture required. If a difference of 1% or 2% in remaining moisture is of no consequence, then the answer is Yes, because the liquor builds up in the outer case and does not come out immediately it leaves the basket. On the other hand, the last traces of moisture do not always show very clearly in the amount of the drippings and may be picked up by the air which circulates between the basket and the outer casing. The best plan is undoubtedly to run for the time which trial has shown gives satisfactory results, and not to give undue attention to the drippings.

Great attention is often paid to the amount of centrifugal force employed. This is usually not as important as it is supposed to be and is certainly subject to a law of diminishing returns. When once a certain speed or effect has been attained dependent on the product being spun, further increases will add a great deal to the power required and to the general wear and tear, but very little to the results in regard to dryness and speed of operation. This is particularly the case for large open crystals, and in some measure for textiles and many other substances. Tests made by the writer on various materials showed that for thin layers the relationship was much as follows :—

$$R \log C = K.$$

Here $R =$ the amount of remaining liquor per

hundred lb. of dry substance, C is the centrifugal effect, K is a constant.

It is not suggested that this formula holds exactly on a large scale, but it gives results not widely out of touch with much practical experience. For most purposes the speeds given in Fig. 4 are more than sufficient, and, in fact, machines running at 10% or 15% less than those indicated therein will give quite tolerable results. For laundry work a 26-in. machine running anywhere at 1,200 r.p.m. or 1,300 r.p.m. will do very useful work and the effect of speeding up even to 1,500 r.p.m. will not be very marked. In handling certain slurries it is necessary to use a comparatively low effect during the filling operation, since a high one will cause heavy solid matter to pack at once on to the rim of the machine and in many instances it becomes so impermeable that the liquor is hindered in its passage and in extreme cases cannot get away at all. This effect may be especially marked in difficult filtrations if a large body of liquor is allowed to build up in the machine, as then the liquid pressure increases the packing of the solid portions. The best way in such cases is to feed no faster than the liquor can escape. As soon, however, as the bulk has drained away the machine may be brought to full speed in order to complete the final drying.

The thickness of layer plays a very small part when materials are of an open texture, but for close-textured materials thick layers require more time both for filling and for subsequent drying.

It is, however, rarely necessary to limit the thickness except for very slimy products or for purposes of washing as in the case of some sugars. The layer next to the basket usually retains the most moisture, because all the drainings pass through it, and probably on account of capillary effects also. The inner face is usually the dryest, and in a test on salt the moisture increased steadily from top to bottom of the basket, owing probably to the increasing thickness of layer. A short run with caustic mud showed the wettest layer to be intermediate, but on longer runs the rule was followed. In textiles the dryest layer is sometimes intermediate.

We can now consider the question of loading the machine, and the first question is that of capacity. In laundry work the machine is usually packed to its limit and a stout cloth is placed over the goods, and pushed under the rim of the basket, to prevent any work flying out and as protection against dirt and fog. There is often a tendency to waste valuable time in forcing in a final small amount of linen, and it is more economical not to try to load above the level of the basket rim. As soon as the machine has come up to speed the clothes will be found to have packed themselves and will occupy approximately the net capacity measured by the volume under the rim. Yarns and such materials are generally laid round the rim and occupy something less than the net capacity. Muds and so forth can be charged into the basket to an extent governed by its net capacity, but many crystals, even when fed in

the form of a slurry, will occupy a much greater
volume and sometimes may reach as much as
60% to 70% of the total capacity, depending on
the angle of repose, the depth of the basket and
the manner of the filling. As a guide to capacity,
it may be said that laundry goods can be charged
into the machine to an extent of 12 lb. or 15 lb.
of dry fibre per cubic foot of gross capacity. As a
guide to the meaning of this as regards the number
of articles, it may be said that sheets weigh from
1 lb. up to 3 lb. apiece, an average being, say,
1½ lb. in the dry state. Caustic mud contains
about 60 lb. of dry material per cubic foot occupied,
sugar may be taken at 51 lb., vacuum salt 56 lb.
and sulphate of ammonia 56 lb. All these figures
exclude residual moisture. It is usually not a
difficult matter to estimate the weight of a cubic
foot of any material in case of need by packing
it firmly into any receptacle of known volume and
then weighing it, and a little experience is a very
good guide to the extent of load which could be
got into a basket of given size.

In order to get even loading, various methods
are adopted. In laundry work, the simplest
method is to fill the basket right to the limit of
its gross capacity. The correct method is to put
in small compact amounts of material, spacing
them regularly round the basket and pressing
them down evenly as filling proceeds. By this
means any one article is kept in a position to
itself. If this precaution is not observed a part
of a sheet, for instance, may be caught by portions
in another part of the basket and tearing will

occur as soon as the speed rises and the goods become compressed.

When the basket is filled while at a standstill, care should be taken to see that the surface is evened off so that the depth is equal all over and any necessary spading should be done to see that the consistency is also equal. Slurries are usually fed into the basket while the latter is in motion. Free liquor should not be allowed to collect if the machine is of the self-balancing type (see the next chapter), or surging may occur. If it is likely to be present, care should be taken to see that the speed is well above the critical speed. If the products are of a fairly solid form and are fed into the basket while in motion, a slower speed is often found preferable during the filling in process as in the case of a self-balancing machine under the critical speed the lighter portion is nearer the centre and will tend to pick up a greater quantity of the incoming products. The opposite occurs above the critical speed and while with most materials the point is not of great importance the writer has known such a machine to run increasingly out of truth, due to very heavy solids being run in with the basket running at high speed.

Sometimes a spud or bar is used to even up the load in the basket while the latter is in motion and before bringing it up to full speed, but this operation, though sometimes convenient, cannot be commended from the point of view of safety first.

Washing is often carried out in the centrifugal

and in some cases is very economical. It is often done by means of a bucket, watering-can or hose, but the best results can only be obtained by the use of properly designed nozzles giving a fine soft spray. Coarse sprays which flush the surface of a load with free liquid are almost bound to cause cutting or channelling, and if the surface of the load is somewhat sharply angled it will be almost impossible under such conditions to prevent the liquid flowing unevenly. A definite quantity of wash liquor should be used for each load and apparatus is available for supplying this. Too great a thickness of load is not always desirable in the case of soluble crystals, as the wash liquor may be saturated by the time it reaches the outer face, which will then not be adequately purified unless an undue amount is dissolved from the inner layers.

Conditions in regard to emptying are very varied; some crystals will fall out of the basket of their own accord immediately it stops, others require a slight touch to bring them down, and still others will have to be dug out with the spud. In some instances, the difference is solely due to increased dryness and packing due to longer running. Some muds pack very firmly and are hard to dig out, and in such cases the best way is to start cutting out, if possible, a portion at the bottom, depriving the wall of material of its support. Alternatively, the wall may be cut out in rough vertical segments.

Fortunately experimental work when required is usually comparatively easy, and a 12-in. basket

2

is extremely useful for handling trial quantities. It should be run at a speed giving the same effect as that of the large machine to be finally used. If the product is difficult to filter or is in the form of a very thin slurry, regard will have to be had to the thickness of the wall, and in difficult cases it is desirable that this should be much the same as that to be finally used. In ordinary cases, experience will allow a correct commercial estimate to be made, even though such factors as wall thickness, etc., differ, but there is no simple proportion which can be generally applied with ease. It should not be forgotten that as a rule it is not convenient to fill a large machine as rapidly as a small one.

It is interesting to compare the results given by a centrifugal dryer with those obtainable by other means. The general rule, of course, holds, for it is usually cheaper within reasonable limits, to remove moisture mechanically in a centrifugal than by any form of heating. For textiles, swarf, sugar and coarse granular products the centrifugal dryer holds its own against all competition. It has very largely ousted the hydraulic press for dealing with fish and meat offal, and in certain cases compares favourably with it for dealing with oil seeds. It is not, however, a serious competitor yet with the hydraulic press for ordinary oil-mill use on a large scale. For substances which will readily form a thick cake, say up to 6 or 8 in., it competes with open filters and even with the filter-press. It has, however, not got sufficient filtering surface to compare with filter-presses on more difficult problems where such thick cakes are not obtainable, particularly if the product is of a slimy nature, which rapidly chokes the filter-cloths. The centrifugal dryer in such cases may give place to the centrifugal subsider, particularly if the amount of solids to be dealt with is small in proportion to the amount of liquor. Caustic mud is an excellent example of the limiting problem, and may be dealt with efficiently either by an ordinary open vacuum filter, by a centrifugal dryer, or centrifugal separator, or by several types of filter-press. The choice in this case is purely an economic one, and may vary according to the conditions of the particular case, the amount to be dealt with and so forth. A substance easier to filter than caustic mud might fairly be looked on as suitable for handling in a centrifugal dryer, while for one more difficult some form of filter-press or subsider would probably be preferable.

CHAPTER III

CENTRIFUGAL dryers are sometimes referred to
as hydro-extractors or hydros, particularly in the
laundry industry, and where the liquid to be
removed is water. The distinction, however, is
not of any great importance. Other terms in use
are "spinner" or "whizzer." The essential
working part of the machine is the basket or cage,
which is either constructed in wire or of perforated
plate. A wire basket is extremely suitable for
textiles and gives the freest possible drainage.
It is satisfactory from the point of view of safety,
since it is not likely to give way without there being
plain warning as to its condition, yet it is not so
strong and suited to rough handling as the plate
type of basket, which is exclusively used for the
handling of various crystals and chemicals, and is
in a fair way to become so for laundry and textile
purposes. As a matter of fact, its efficiency for
the latter use is so little below that of the wire
basket that it is often difficult to measure by any
practical means the difference in remaining moisture
on any operation of reasonable length.

Baskets are made of a very wide variety of
materials, though steel and copper are the most
ordinary ones, and are usually tinned or galvanized.

Phosphor bronze baskets are often used where acid materials have to be handled and are quite commonly installed for sulphate of ammonia; aluminium and monel metal are also occasionally employed, and it is an extremely common practice to coat with rubber or vulcanite. The perforations vary from about $\frac{1}{4}$ in. in diameter down to, say, $\frac{1}{8}$ in., but if fine crystals or slurries are to be treated a liner must be put in. In many cases a plain gauze is sufficient, but very often a filter-cloth is required. This is preferably supported upon gauze, in order to allow free drainage, and in some cases it is protected by a coarse gauze laid over it. Alternative methods sometimes used are to enclose the material to be spun in canvas bags, which are packed into the basket in the ordinary way, or a single bag may be made to fit neatly into the basket, its edge being clipped to the basket rim.

The size of basket ranges from, say, 4 or 6 in. for laboratory use up to 84 in.—a size which is capable of handling such problems as large carpets weighing a ton or more when wet.

Baskets may be further classified according to their discharge arrangements. Plain baskets are, of course, in vogue for textile materials, laundry and other purposes. For laundry use favourite sizes are 26 in. and 30 in., although a good many larger machines are to be seen; 48 in. is perhaps the largest size in any general degree of use. They are also employed for various chemical purposes, for handling crystals such as soda, sulphate of ammonia and countless other materials, particularly in instances where only one or two units are required.

Sometimes two or more doors are arranged in the bottom of the basket, and at the close of the operation these can be opened and the dried products pushed through on to a shoot arranged in the outer casing. Generally it is necessary to dig

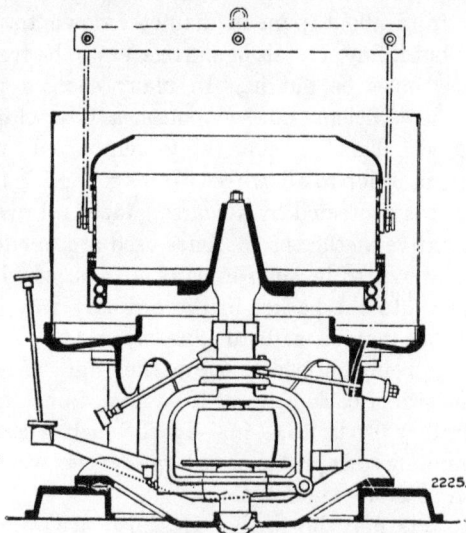

FIG. 5.—Lifting-type Basket.

Fitted with mercury balancing rings, in free spindle centrifugal.

away some of the material in order to get at the doors, and while they present some advantages, they do not compare in convenience with the types described below, particularly when a battery of machines is to be installed.

Baskets of the lifting type, such as are shown in Fig. 5, are often met with. They are particularly in use for handing metallic swarf, fish and meat offal, etc., and many such machines are to be found in sulphate of ammonia plant. Often the whole basket is arranged to lift out and tip, but in other cases there is a false inner cage or lining which is lifted out by means of a small jib crane arranged at the side of the machine. The inner cage is generally fitted with a loose top ring, which can be unclipped to facilitate the removal of the goods. Capacity in such cases is much increased by the use of two baskets, as one of these can be unloaded and reloaded while the other is being spun.

A very favourite size for handling meat offal is 36 in. while 40 in. and 48 in. are in common use for crystals. For swarf, smaller sizes are ordinarily in use, ranging from 14 in. to 20 in. It is common to to put several of the smallest units at convenient positions in a machine shop, so that the labour of carrying swarf to them is minimized. Where the swarf is dealt with in bulk, however, larger units up to 40 in. or 48 in. are installed.

The centre-discharge type of basket is in very common use for handling all types of crystals and sludges, and is shown in Fig. 6. It is mostly installed in batteries, but single units are often seen. The basket is attached to its spindle by means of a spider, which forms a discharge opening, covered by means of a light valve. At the completion of the operation, the valve is pulled up and caught by a hook on the upper part of the machine and the material is spaded down. Soda crystals and many others, which build up over the central valve, will fall down of their own accord if the

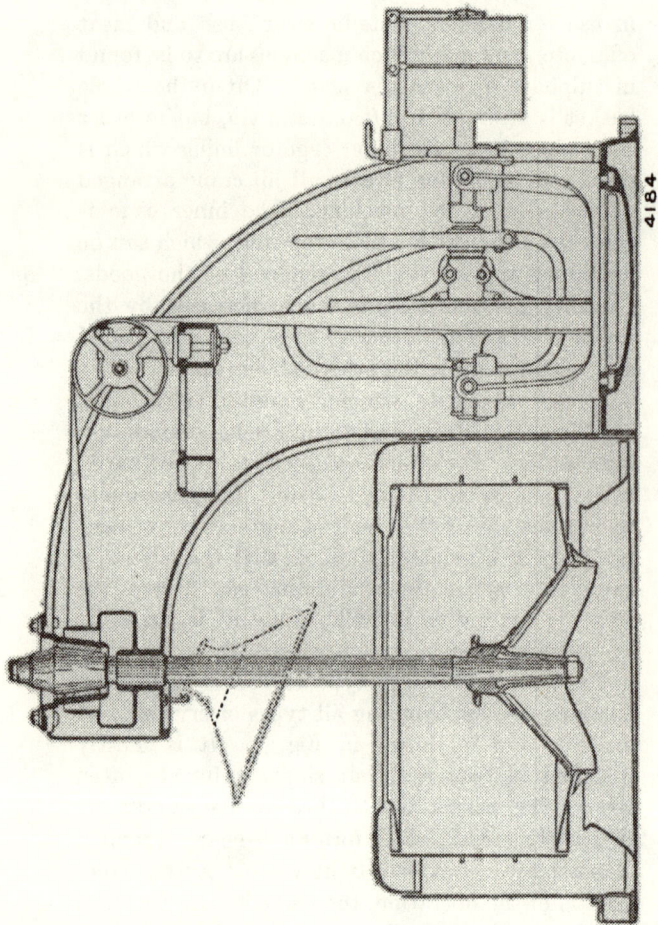

4184

Fig. 6.—Weston Centrifugal—with Central Discharge.

latter is raised or lowered sharply once or twice, by inserting a lever under its upper flange. Machines of this type are found in all sizes from 26 in. upwards; 36 in. and 42 in. are very common, and probably the favourite unit for large installations is 48 in. A modification forms the self-discharge pattern. In this case the central opening is not covered, but a cone-shaped distributor, not unlike the valve, is fixed on the shaft about half-way up the basket. The incoming material is run on to this cone, which throws it to the wall of the basket, and the centrifugal force prevents it from falling out, so long as the basket rotates. On stopping the basket, the spun material, if of a loosely packing nature, slides through the central opening of its own accord. This type, however, can only be employed for materials which flow into the basket in a plastic condition, building up almost a vertical wall of dried product. This is the case with certain classes of sugar, but granulated and crystal sugars are more generally dealt with in the central-discharge type, as they are spun dryer, and pack harder.

For freely discharging materials which do not tend to form a vertical wall, but build inwards at the bottom of the basket, a central valve may be used, elongated by a cylindrical lower portion so that the top comes near the upper part of the basket, where it is controlled by a forked lever, by which means it can be lifted.

Scrapers such as are shown in Fig. 7 are used, particularly in America, in cases where the wall of crystals does not tend to fall down of its own

2*

accord. At the end of the drying operation the basket is run at a slow speed, and the scraper, which is mounted on the outer casing, is swung over and caused to traverse up and down the wall

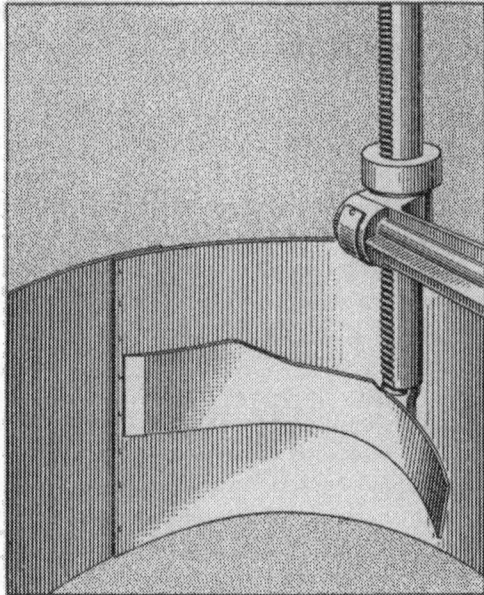

Fig. 7.—Roberts Gibson Discharger.

of material by means of a hand-controlled rack motion. This is said to save a considerable amount of labour, but it would appear necessary to employ some care in its operation, and it may be somewhat hard upon the linings.

Innumerable special baskets have been made, with compartments for taking trays, reels, bobbins, boxes and so forth, but as a general rule the simpler the basket construction the more likely it is to give satisfaction in practice.

Baskets for fixed-spindle machines usually have a brake-ring mounted on the bottom. This permits a large brake effort, and no strain is thrown on the spindle. The band is supported on the bottom of the outer casing and carries a number of wood friction blocks, which are often faced with ferodo. Where this is unsuitable on account of corrosion or fire danger a brake-drum is applied to the spindle, and this is of necessity the practice in machines of the self-balancing type. Here, again, a steel band with ferodo lining is in extensive use, although a single plain wood shoe is often seen. In general, the pressure has to be maintained by hand, but in some types a clip holds the brake in position, under suitable tension.

The outer casing is ordinarily constructed of cast iron in the case of fixed-spindle machines, and of steel in the case of the self-balancing type, as the basket may strike the side in extreme cases of bad loading. Neither should be regarded as a safeguard against bursting, which is an extremely rare occurrence provided that the basket is given a reasonable inspection from time to time, unless it is worked under very improper conditions as to speed or loading. Another cause may be unsatisfactory repairs, carried out by a firm lacking experience in this special class of work.

Outer casings or pans are often lined with lead or vulcanite. Occasionally, where they are to be used for foodstuffs, and particularly cleanly conditions must be observed, cleaning doors are arranged to give direct access to the interior. In other instances, the top and sides of the outer casing may be entirely removed from the base, to which they are jointed by a rubber ring and suitable screw clips. This gives ready access in cases where clogging of the outer casing may occur through crystallisation, but this is rarely required in ordinary practice.

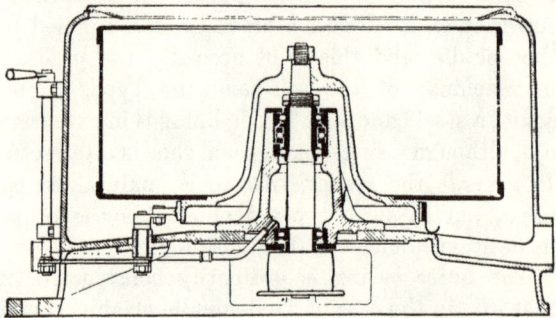

FIG. 8.—Fixed-spindle Centrifugal.

Centrifugal machines generally may be either of the fixed-spindle type, or of the free-spindle pattern, which is generally referred to as " self-balancing." This latter term is somewhat of a misnomer, as it is quite certain that the self-balancing machine requires an adequate degree of care in loading, or it may refuse to run at all. In fixed-spindle machines the bearings are rigidly fixed to the base or supporting arms (Fig. 8), and the spindle is only free to rotate and cannot give to any unbalanced load. This therefore is thrown entirely on the bearings, which must be generously dimensioned in order to cope with it. It is, however, astonishing to find how excellently such bearings stand up for long periods without any special attention or renewals, even though loading may not be particularly skilful. Nowadays the general practice seems to be to employ roller journal and ball-

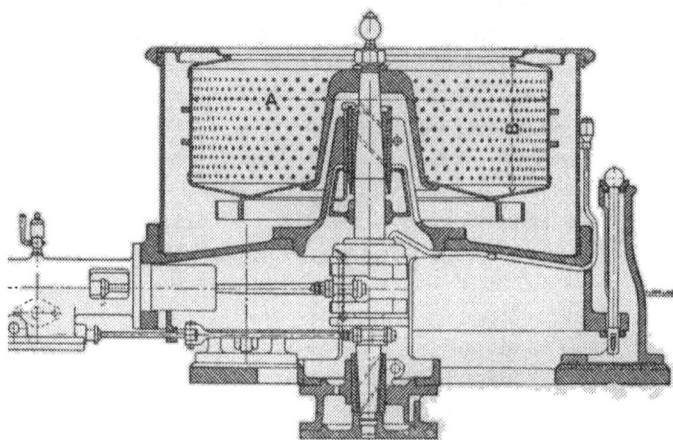

FIG. 9.—Lubrication System, Direct Steam-driven Centrifugal.

thrust bearings, and the grease in these requires to be renewed every six to twelve months. It is a good plan not to rely entirely on the grease gun or stauffer, but to take the bearing apart and clean it out from time to time, to ensure that no metallic grit accumulates.

Plain bearings are, however, still in use; and Fig. 9 shows a good system applied to a direct engine-driven machine. The upper and lower bearings are tapered and their height can be adjusted to take up wear. Oil is supplied to the top of the upper bearing, and circulates through it, being caught at the lower end by a cup attached to the spindle. As oil accumulates in the cup, it builds up sufficient pressure under centrifugal force to cause it to return, through a stationary pipe, to the upper part of the bearing.

The crank-pin is lubricated by a cup or ring mounted above it. A trap is arranged in this in the form of a U-pipe, into which a certain amount of oil is forced at each operation. As the machine slows, this is free to descend to the crank-pin. A portion, however, flows down another channel to lubricate the bottom footstep. This may consist of a phosphor bronze plug in the end of the shaft, resting on a steel toe-piece, or alternate steel and bronze discs are sometimes used.

The advantages of the plain fixed-spindle machine are its simplicity of operation and construction and its general reliability. It can be loaded and run by labour of low intelligence—for instance, in a workhouse—which is not capable of handling the self-balancing type. Its disadvantage is that bad loading may cause vibration to be transmitted to the foundation, and in some cases this may be undesirable. This may be overcome by suspending the machine by means of rods and pillars carried by an external base-ring. The rods are attached by hemispherical joints to the pillars and outer casing, so the whole machine is free to swing and no vibration is transmitted. It then becomes suitable for mounting on upper floors, or the supports may be made of such a height that it is situated conveniently for discharging, by means of a shoot, into a barrel, as shown in Fig. 10. Sometimes the suspension rods are screwed and may

FIG. 10.—Suspended Machine, mounted on Pillars, for Bottom Discharge.

be used as screw-jacks for raising the machine, permitting easier access to motor, steam engine, or bottom bearings. Although this type is free to swing, yet it will be found remarkably steady under bad loading, since this is opposed to the inertia both of basket and outer case, and the critical speed is fairly low. Examples will be found in use for all types of work, both laundry, textile and chemical.

In the self-balancing or free-bearing type of centrifugal, the spindle bearing is carried by a flexible support or buffer. This allows the basket to find its own centre of rotation, which at high speeds usually corresponds with its centre of gravity. The chief advantage of this construction

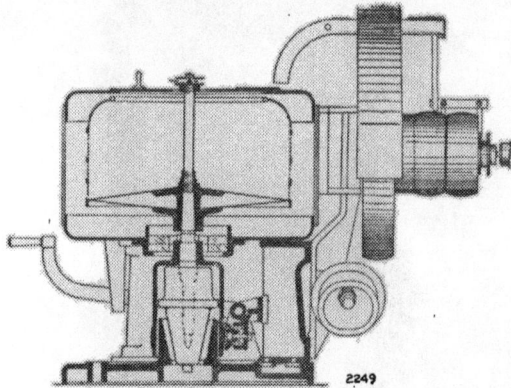

FIG. 11.—Self-balancing Centrifugal. Under-driven.

is that side-stress on the bearings is limited, which is essential if extremely high speeds are required. It is a very useful construction for top-driven machines, as the absence of a bottom bearing permits a very convenient arrangement of central discharge. A certain degree of care is required in loading, but the operator soon learns that bad loading leads to a waste of time and trouble. There are many types of such self-balancing machines. Under-driven machines up to 36 in., as shown in

Fig. 11, are largely in use in laundries. For other purposes machines are built up to 48 in., but in these larger sizes it is customary to use an arrangement of rods controlled by spring and rubber buffers at their outer ends, and attached at their inner ends to a pivoted bracket, which carries the bearings. An example will be found above in Fig. 5. One of the best buffers for the smaller sizes is a rubber one of conoidal form, which is placed in a spigot housing, and into which the bearing box is dropped. Its success depends largely on its taper being correct and an angle of not more than 20° is usual. It is self-adjusting and provides a good deal of internal friction, which, as we see later, is desirable. This buffer is also in extremely common use for top-driven machines of the Weston type, of which one will be seen in Fig. 6. This class of machine is in almost exclusive use for handling sugar crystals, and is extremely common for other purposes where a bottom discharge is requisite. It is also suitable for laundry and textile use where self-balancing machines of 36 in. diameter and over are required. The upper bearing is often supported by swan-neck brackets, but for direct motor or pelton wheel drive, or large groups of machines, a girder arrangement is usual.

In a modified type of the Weston machine shown in Fig. 12 a rubber buffer is not used, but a spherical upper bearing is employed. This serves merely to position the spindle, and takes no weight. In order to provide the requisite amount of friction and restoring effort, the weight of the spindle is carried on a loose bottom bearing, which is free to

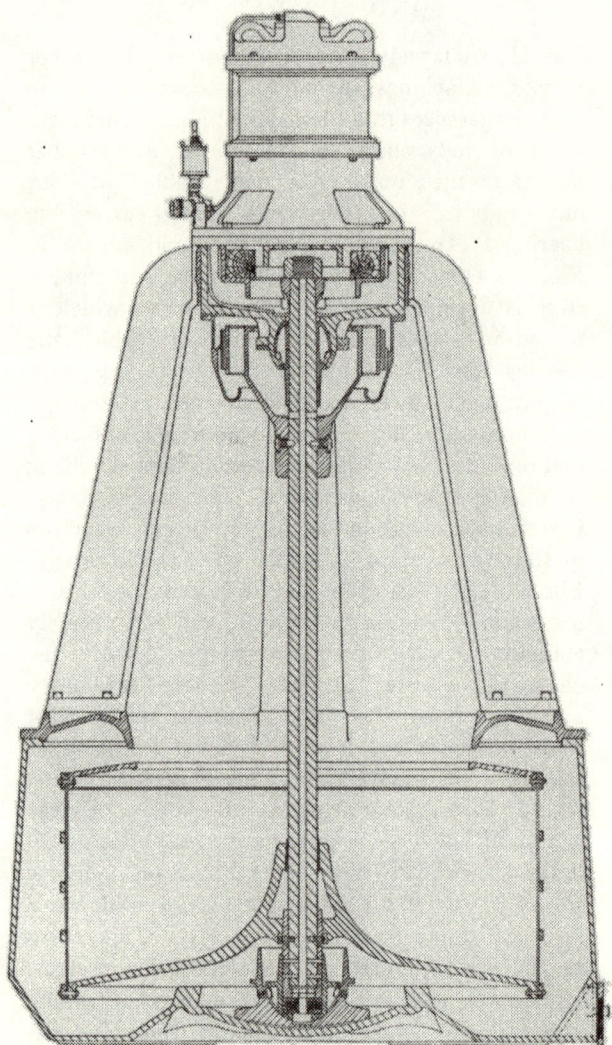

Fig. 12.—Self-balancing Centrifugal with Friction Shoe.

50

slide in a cavity in the base plate shaped like the segment of a sphere. The friction fixes the basket until the critical speed is passed, and the method gives a very satisfactory control.

Many forms of adjustable buffer are in use. A common plan is to arrange a flange on the bearing-box and place rubber rings above and below this. The whole is dropped into a suitable spigot and screwed or clamped down. For most purposes, however, the self-adjusting type is preferable as the operator will nearly always overdo the adjustment in an endeavour to correct errors due to bad loading or other causes. This is apt to intensify the trouble and cause worse vibration.

Free-spindle machines are subject to two types of disturbance, of which the first or whirling vibration is noticeable during a certain critical period of the acceleration, and arises from faulty balancing of the load, or a bent spindle. If both these matters are in reasonable order and trouble still occurs, it will generally be found that the buffer is too tight, or is so worn that the proper degree of friction is not present. Another contributory cause may be poor acceleration.

The other trouble is a slow gyration which is usually most prominent at the higher speeds, although it may occur at almost any speed. This is known as precession, and may be due to air swirl between the basket and outer casing, which can usually be cured by fitting baffles inside the latter. Another cause may be undue friction in the universal joint, if present, which makes the connection between the motor or water turbine shaft and the basket spindle, or the buffer may be too slackly adjusted, or worn out. External vibrations of the correct periodicity, particularly in the transmission, may occasionally contribute to either trouble.

A rough test for a satisfactory buffer is to push the spindle to one side and note how rapidly it comes to rest. Unless the deflection is very great

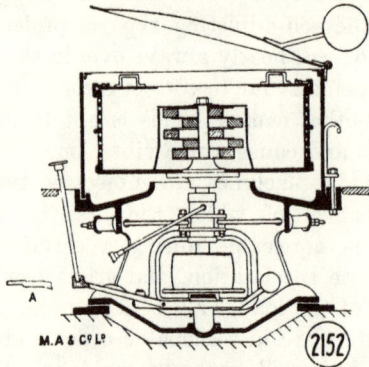

FIG. 13.—Compensating Rings Fitted to Self-balancing (U.D.) Centrifugal.

it should do little more than swing once past the centre and then immediately settle down.

In order to make the machines truly self-balancing, automatic compensating devices are sometimes fitted. In Fig. 5 above, the basket is shown fitted with a couple of circular tubes containing mercury, which are attached to its under side. Above a certain critical speed the basket tends to rotate about a centre which is nearest its heavier side and the mercury flies out to the opposite side and provides compensation within limits. In another device, Fig. 13, a number of rings are fitted loosely around the central spindle, inside a dome in the interior of the basket. These rings fly out and tend to set themselves towards the lighter side of the basket above the critical speed. A difficulty with all these devices is that if they come into operation at all at low speeds they tend to act with the unbalanced load until the critical speed is reached, but they contribute considerably to smooth running when this point has once been passed. They are also liable to hunt, and for these reasons and the additional cost entailed, they are not at present in any very wide use.

It is not necessary here to elaborate greatly on the means of driving. Belt-driven machines are much in use, and are low in first cost. It is usually necessary to provide a counter-shaft in order to obtain the required speed. This may form part of the machine itself as shown in Fig 11 above. Sometimes a cross-shaft is mounted on a top arm above the pan and the drive is transmitted to the spindle through bevel friction wheels, one of which has a leather or papier mâché face, while the other is of steel. The friction is controlled by a spring pressing against the end of the cross shaft, and the machine is started and stopped by a lever controlling this spring. In all cases, where belts are used, it is desirable to shield them, if possible, from any contact with stray liquor or vapour which may emanate from the material being treated, particularly if this be corrosive.

Steam drive is useful for machines in isolated positions not accessible to belt drive. As we have seen, the spindle may actually form the crank shaft of the steam engine and the cylinder is then mounted on the side of the pan, as in Fig. 9 above. In spite of

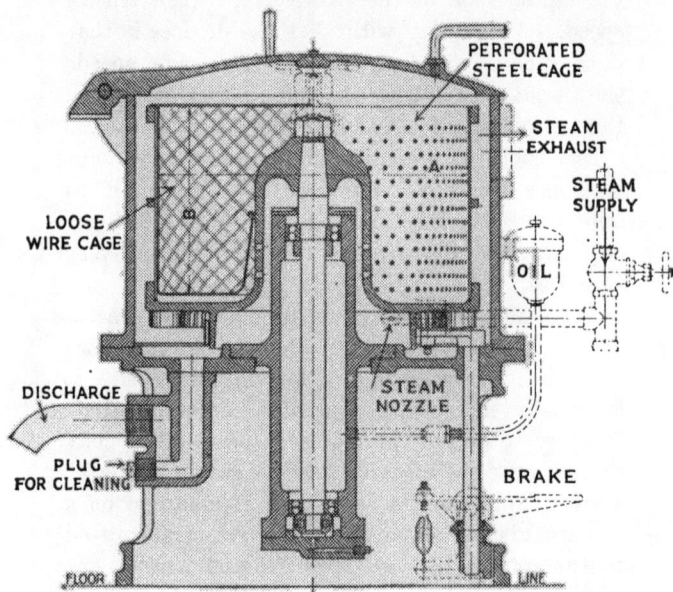

FIG. 14.—Turbine-driven Centrifugal.

the high speed, these engines stand up remarkably well and the steam power consumed is not excessive. Nevertheless, a direct drive of this nature is hard on the reciprocating parts, and the tendency is to use other types. A separate steam engine is often mounted on the base plate of the machine, driving through a belt, and forms a satisfactory solution for many requirements. Turbine-driven machines are used, but are steam consumers and are only economical where the steam is of use for treating the material in the basket, or for the very small high-speed separators described in a later chapter.

A typical example is employed for treating greasy rags and similar substances and is shown in Fig. 14. Here the turbine blades are mounted on the bottom of the basket and the steam is guided inwards through suitable perforations in the centre dome into the basket; the outer casing in this instance must be fitted with an exhaust pipe.

Electric drive is clean, compact and economical in power consumption, particularly when the drive is of the direct type. Labour in starting and stopping the machine is small and it is comparatively easy to arrange devices for automatic operation and guarding. Fig. 15 shows a special type of electrically-driven centrifugal which is so arranged that it can be set to run for a definite time—5, 10 or 15 minutes. At the end of this period the machine brings itself to a standstill, and switches over from a white to a red light, indicating that it is ready for discharge. Means are also provided for locking guards in position while the machine is running. This particular machine is interesting, since the shaft is flexibly supported at its upper end, and rests in a flexible bearing in the outer casing, which is free to swing, thus combining the principles of the Weston and the suspended machines.

Water-driven centrifugals present many of the advantages of the electrically-driven machine when

FIG. 15.—Electrically Driven Centrifugal, Automatic Type.

arranged in a battery so that a number can be run from the one pump, which is usually of a duplex type.   By this means the trouble and complication

of belts and shafting are avoided. The turbine
is of the Pelton wheel type (Fig. 16), and is provided

FIG. 16.—Pelton Wheel, Flexible Joint, and Buffer
for Water-driven Centrifugal.

with two jets, both of which are in use for accelerating, but one only for running. The second jet is
shut off when speed is reached either by the use of

a governor or by a hydraulically operated valve, which may be set in motion by the overflow of water into a control chamber. This overflow is arranged so that it only occurs when the machine has reached such a speed that the discharge from the turbine blades begins to travel in the same direction as the pelton wheel.

Friction clutches are much in use to enable the drive to be taken up without shock and without making undue calls upon the driving mechanism or electric motor. They may be of the hand-operated type, such as is seen in Fig. 6 above, and this is extremely convenient if it is wished to run the machine at a slow speed during filling, as the clutch can easily be manipulated to this end.

Centrifugal clutches are also used and their operation will be described in a later chapter. These take up the drive very gradually and are particularly requisite for motor drive. They do not transmit the full power until the motor has gained sufficient speed and can be so constructed that no power at all is given out till quite a high speed is reached. This enables the use of motors having a low starting torque.

Washing and steaming devices are in common use and in their simplest form consist of a hinged pipe carrying one or more spray nozzles, which can be swung into position when required. A cover is often fitted in addition. Such an arrangement may be elaborated by the use of a regulating cylinder, which delivers a given quantity of wash liquor per operation. If it is particularly desired to separate the washings and the original mother-liquor, this can be done, but a somewhat special basket and outer casing are necessary. An outer imperforate shield is attached to the basket proper and receives the liquors, which are discharged from its lower rim. The outer casing is divided into an upper and lower compartment and a ring is fitted which can be drawn up, when required, above the liquor discharge point to divert the flow.

If it is desired to dry products which are very permeable, the machine can be fitted with a cover, and hot air forced into the basket by connecting up a fan and heater. This has been applied for drying small metal articles and fittings, and similar problems.

Machines may be made self-ventilating by connecting up a pipe of sufficiently large diameter. This should enter the casing at a point as remote as possible from the centre, and on occasion it is mounted on the drain. The fan action of the basket will force the fumes up this and cause air to be drawn in at other openings provided the general design is suitable. This air does not necessarily go through the basket, but may pass round it. If there is a large bottom opening and a somewhat larger one in the top ring a certain amount of air is liable to be blown out of the top of the machine, and this may carry liquor or spray with it.

The question of guards is becoming of increasing importance. Operators are often liable to do foolish things, such as putting the hand on the basket or throwing extra material (such as linen

or other products unsuitable for such treatment)
into it while in motion. This behaviour is apt to
lead to serious injury, and from time to time fatal
accidents are produced. Large numbers of
machines are unguarded, but it is hardly likely that
such a state of affairs will be permitted to exist
much longer. There is no necessity to elaborate
here the methods of guarding the gearing; these
will be obvious. Hinged guards fitting on the
pan are quite simply arranged, but should be of
open mesh in the case of under-driven self-balancing
machines, so that it may be seen whether the
basket is running properly. It is quite easy to
link up the cover with the starting gear, so that
it must be closed before the machine can be put
into motion. Connections with the brake can
also be arranged quite simply, and the guard can
easily be kept locked in position till the current is
switched off (or the belt on the loose pulley) and
the brake applied. It is a less simple matter to
arrange that the guard cannot be opened until the
machine is stopped or nearly so. The difficulties
are mainly those of the additional expense and
trouble in operating, and the fact that many
devices readily get out of order and are not sufficiently simple and positive. In one device, a
hinged blade is hung between the basket and
outer casing, and the swirl of air, when the basket
is moving, deflects this and causes it to operate a
catch securing the cover. In another device two
lugs are arranged on the driving pulley, and these
displace a lever which locks the cover quite
effectively. It is impossible to replace this lever

in the opening position until the speed has been reduced almost to zero, and the basket must be stopped with the lugs disengaged. This is easily accomplished by a manipulation of the

FIG. 17.—Automatic Guard.

brake, for it is quite possible, when the basket has almost stopped, to suddenly lock it to an indicating mark.

In another device, shown in Fig. 17, the spindle of the basket carries a brass ring, which is fitted into a circular casing of similar diameter. The ring, however, is free to fly out a certain distance under the action of centrifugal force, and locks the basket cover in position. Before the cover can be raised the speed of the machine must be reduced to such an extent that the ring can be pushed back into the casing by hand.

Numbers of special machines have been put on the market, but comparatively few of them are found in extensive use, and the principles of operation will be quite readily understood from the foregoing descriptions. Usually the modifications apply to the basket and consist of special fittings to take trays, reels of piece goods or other special apparatus.

Numbers of special nitrating centrifugals have been made and the more elaborate types form a good illustration of specialised plant. Two speeds are provided, of which one is very slow, for circulating the mixed acids, and the other a fast one for removing them from the nitrocellulose. Basket and fume-tight cover are usually constructed of aluminium and the outer casing is provided with a cooling jacket and fume pipe. A hydraulic conveyor is a very useful adjunct, and the head of this is sometimes practically integral with the machine. It consists of a chamber into which the spun nitro-cotton may be pushed direct from the basket. In this chamber it is taken charge of by a water-spray, which forces it into the conveying pipe, and any exposure to air is reduced to a minimum.

Machines with horizontal spindles are very rarely seen, but one interesting application is treating certain textile materials when wound on a spindle after dyeing. In this case, two horizontal spindles are employed, each of the self-centring type. One of these may be drawn back as required, to allow the spindle carrying the yarn to be fitted in, and a loose outer casing is then lowered and the machine operates in the ordinary way. Such a device is particularly useful where relatively small quantities of material have to be treated, and various dyes are in use. Since the thread does not come in contact with anything except its own spindle, there is no danger of any contamination by an alien colour.

Before we discuss continuous centrifugals, it may be well to form some idea of what can be done with ordinary batch types. We do not propose to give any elaborate figures as to loads in baskets of any given size, since the capacity will vary with the make and type, *i.e.*, self-balancing machines usually have a somewhat greater capacity for a given diameter than the fixed-spindle type, since it is usually practicable to make them deeper. The power required also varies with the means of driving, but the following figures may give some idea of what should be provided for steady running :—

26 in. machine	$1\frac{1}{2}$ h.p.
36 in. ,, ...	$2\frac{1}{4}$,,
48 in. ,, ...	$4\frac{1}{2}$,,
60 in. ,, ...	$6\frac{1}{2}$,,

For purposes of acceleration, it would be usual to provide something like double the above figures.

The actual internal friction in the machine itself, apart from any mechanical loss in counter-gear, motor, etc., may be very much less than this, and some trials have shown that it may not vary greatly from 2 h.p. for a 48-in. Weston machine during steady running.

In ordinary use in a laundry, machines are seldom worked to capacity, since the hydro-man may also look after washing machines and will probably have to bring the work to and from the hydro-extractor. One man can work two to three machines, and on sheets and other non-starched work three to five operations per hour can be made, the latter being extremely good practice with a small machine—say up to 30 in. diameter. It is not likely to be seen except where the spinning time is kept as short as possible, and loading is not meticulously careful.

On starched work two or three operations would be usual, while four would be very rapid practice. A 30-in. machine should hold some 60–75 lb. of dry fibre—say 40–50 sheets. A-48 in. machine will hold something like three times the above, but of course loading and unloading will represent a greater proportion of the operation. On other textiles, such as yarns, five or six operations would be quite usual and ten feasible when required, except with the very largest machines.

In handling sulphate of ammonia, or similar salts, which build up very fully, a 40-in. basket should take about 5 cwt. and a 48-in. 7 or 8 cwt. Two and a half to five operations per hour should be averaged over the whole week, depending on the speed with

which the product flows into the basket and the
amount of time taken up in neutralisation, washing,
etc. A sulphate of ammonia machine will usually
have its own attendant, who also looks after the
saturator, and it would be very usual to take 3
minutes and upwards for charging, and $1\frac{1}{2}$ to 4
minutes for discharging through a central discharge.
In handling material like caustic mud, a 48-in.
centre discharge basket will hold about 8 cwt.
of mud having 30% moisture and two operations
per hour would be reasonable. Seven to 8 minutes
might be taken to charge the basket and 6 to 7 to
empty the bottom discharge machine. In a large
plant handling this product two men would operate
six 48-in. bottom discharge machines and another
would be required to look after the pumps, tipping
trucks and agitators. A much larger number of
operations would be got through in the case of
some raw sugars handled in the self-discharging
type amounting to, say, ten or twelve per hour.
Charging might take $\frac{1}{2}$ a minute to a minute and
a similar time for stopping and emptying. One
man would operate two or three machines. Granulated sugar would be dried more fully and would
require more digging down. For this purpose
six central discharge-valve machines might require
a gang of five men, one charging, one liquoring,
and three digging down. Here filling might take
2 minutes, and discharging and washing out
6 minutes.

In handling meat offal in a 36-in. turbine machine
with a lifting basket, two to two and a half charges
per hour would be averaged, about 3 cwt. of pro-

3

ducts would be put into the machine, and a not unusual result might be a spun product containing 30% moisture and 10% fat. The steam in this case might amount to 200 lb. per hour during actual running, though statements are made claiming a smaller consumption. One man should readily operate two or more machines if required and fulfil a certain amount of other duties in addition. Two to 3 minutes are needed for removing a spun basket and replacing it with one containing fresh product. Emptying and re-charging will take 6 or 7 minutes.

We may now proceed to consider the question of self-discharging and continuous machines. The first great difficulty to be contended with is that if we assume a centrifugal effort of 400 to be in use, the average material in the rotating basket will be subject to a centrifugal effort of something like 20 tons per cubic foot, and unless the design is particularly good, this factor alone calls for great strength in all working parts, and will soon give rise to extensive wear and tear owing to the extreme frictional resistance which even a small charge opposes to motion. It is possible to instal a device in connection with the ordinary type of centrifugal which will allow certain materials to be thrown over the side of the basket by a suitably arranged scraper without slowing the machine. So rapid is the effect that a largish basket may be completely emptied in the course of a quarter of a minute. Whether, however, such an apparatus will prove a commercial success still remains to be seen. An automatic self-discharging machine was invented by Sturgeon, and consists, in its essence, of a basket, of which the periphery can slide up and down with regard to the spindle. This is shown more particularly in Fig. 18. The material to be treated is fed down the hollow centre of the

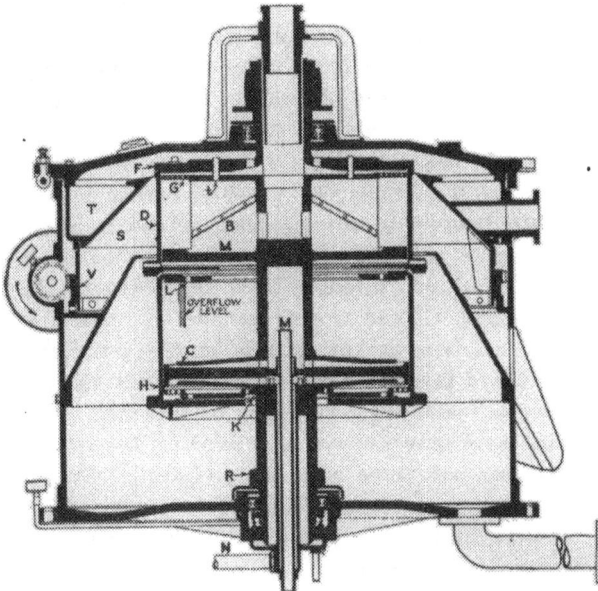

Fig. 18.—Sturgeon Self-discharging Centrifugal.

upper spindle, and enters the basket, B. The solid matter tends to separate out on the periphery, D, and the supernatant liquor escapes through the opening, t, and into the compartment, T, of the outer casing. After a given period the feed is shut off mechanically, water is admitted through the passage, N, and passes to the lower side of the diaphragm, C. It is now retained by the basket bottom, and builds up a wall exerting a centrifugal pressure, which causes the periphery, D, to descend. As soon as movement occurs, a joint is broken at F, allowing any residual supernatant liquor to escape through the filter element, G. As further motion takes place, the solid portions escape into the compartment, S, of the outer casing, where they are carried to the discharge by the rotating scraper, V.

As the basket reaches the bottom of its stroke, the loose diaphragm, K, which is resting on the basket bottom, meets the boss, R, and is lifted, opening a water-discharge port, H. Water is now admitted through pipe M, between the diaphragm, C, and the circular sill, L, which is approximately at the middle of the periphery. The liquid pressure now forces the basket up into position, and the operation is repeated. The diaphragm shown above L is fixed on the spindle and does not move with the periphery. On the downstroke, the liquor below L is forced to the overflow by the fact that the wall of liquor below C is of greater thickness. A controlling mechanism opens and shuts the feed valve at stated intervals, and regulates the admission of the water.

The above machine is, in effect, a combination of the principles of the centrifugal dryer and of the centrifugal separator. It has been experimented with for sewage, and can also handle other materials, such as coal, etc.

Various machines have been made in which the material to be dried is distributed through the central spindle to a number of chambers, arranged round the basket. These chambers are fitted with doors, which open outwards and are operated either by cams or by oil pressure. One or more sides of each chamber is fitted with filtering medium for the escape of liquor. Among this class was the original Schaefer Ter Meer machine, which has found employment abroad for handling sewage, etc.

Among the more fully developed types is the Elmore (Fig. 19), which has been tried for coal, spent sumach waste, crystals, and many varying products in the United States. In this instance, the basket consists of two portions—an inner and an outer, carried on concentric spindles, and driven by gearing which gives a slight differential motion. The outer basket is built up of bars having provision for the escape of the liquid, while the inner is provided with inclined scrapers for helping to move the material. In order to assist the travel, the form of the basket is conoidal, the inclination being towards the discharge level. The material passes in through a distributor at the top, to the space between the two baskets, where it loses its liquor and is helped downwards and outwards, both by centrifugal force and the action of the scrapers. The latter in some cases form little rolls of product which are carried in front of them. The duties claimed are exceedingly high, but very few data are available in this country by which the inventors' claims can be checked back. One would be inclined to suppose that the angle of inclination of the basket might be a point of practical importance, and that the portion which gave good results under certain conditions might not be applicable if these conditions were to vary to some extent.

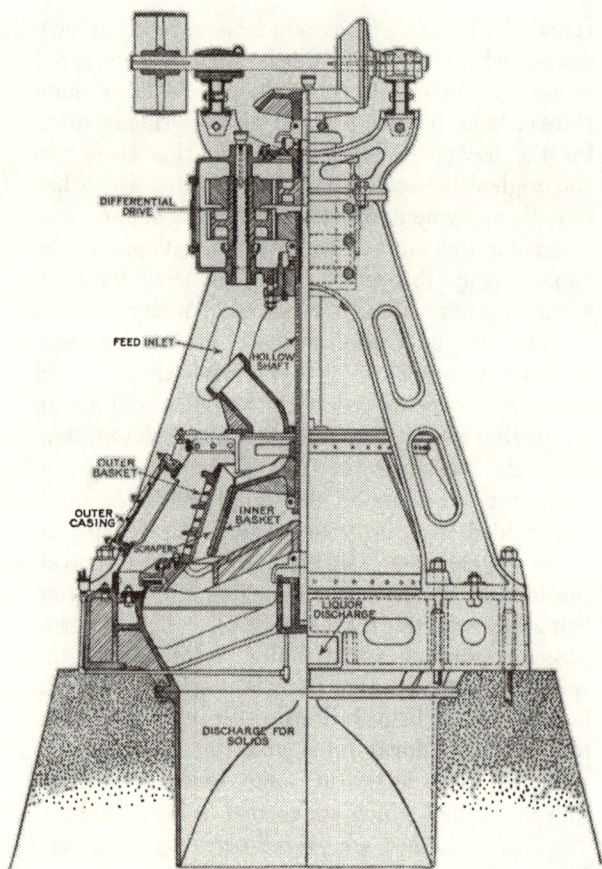

DIFFERENTIAL
DRIVE

FEED INLET

HOLLOW
SHAFT

OUTER
BASKET

OUTER
CASING

INNER
BASKET

SCRAPER

LIQUOR
DISCHARGE

DISCHARGE FOR
SOLIDS

Fig. 19.—Elmore Continuous Centrifugal.

Only a comparatively small amount of material is in the machine at any time, and this reduces resistance and enables higher speeds to be used.

The Hoyle continuous centrifugal dryer also employs an inner and outer basket, but the latter is not coned, and the scrapers form a worm of comparatively small pitch. It is said to deal with 15 to 40 tons of wet coal per hour, for 15 to 25 h.p., the basket diameter being 36 in., and speed 400 to 600 r.p.m.

The Carpenter continuous centrifugal has the merit of simplicity. The basket is approximately conoidal in shape and is formed of a number of screens, separated by saw-tooth baffles (Fig. 20). The material is fed at the top of the basket, drops on to a revolving spreading plate, and is shot against the first screen. The second screen is stepped back a little from the first, and the material, after passing the baffles, is thrown against it with some force. This is repeated from screen to screen till the spun product leaves at the lower edge of the basket, while the liquor passes through the screens to a separate compartment of the outer shell.

The repeated impacts are said to assist centrifugal force in separating liquid and solid. The perforations are $\frac{1}{8}$th in. diameter, but little material passes through, since a layer of it is formed on the screen, and acts as the real filtering medium. Blast-furnace dust has been run through, experimentally, in the form of a slurry with as little as 5% loss.

When operating on coal (washery slurry of a somewhat fine character) a machine with a basket about 7 ft. interior diameter at the base, and 2 ft. deep, running at 350 r.p.m., treats about 40 tons per hour, reducing the moisture from 20%

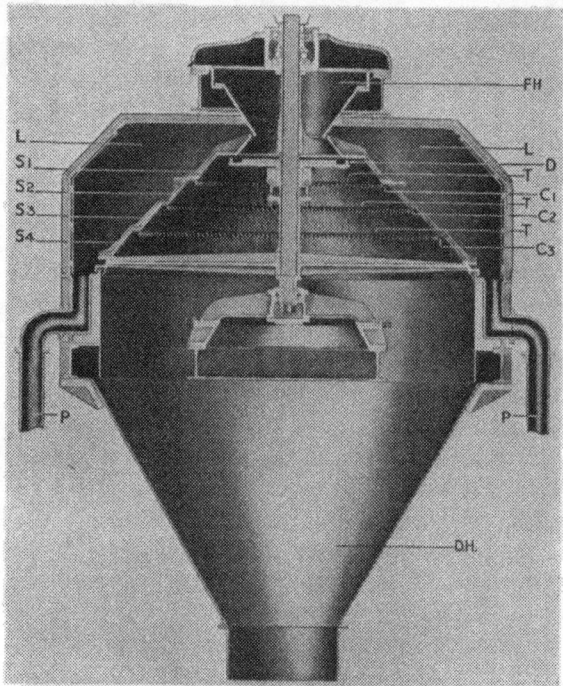

Fig. 20.—Carpenter Continuous Centrifugal.

to 7%. About $3\frac{1}{2}$% of fines pass through, and the power required is about 35 h.p.

Horizontal machines are in use abroad in which the basket has the form of a long cylinder having a ring at one end over which liquor escapes, and at the other a cone piece over which the solids are drawn out by an internal worm-scraper, but the speeds are usually low and the application is more or less limited.

Countless other machines have been proposed. In some the basket is made in halves which are separated from time to time to allow the solids to escape. In the latest device it is proposed to regulate the time of opening by causing the friction of the solids to carry with them a disc as soon as they are built up to a suitable depth. The motion of this disc causes the discharge device to operate.

CHAPTER IV

CENTRIFUGAL SEPARATORS—GENERAL PRINCIPLES

Description of machine ; analogy with tank; type of flow; shape of liquid surface; laws of settlement; aids to separation; diameter of discharge lips— Formulæ relating to milk separators. Power absorbed.

THE machines discussed in the previous chapters had perforated baskets, and the liquor in effect was filtered off. This method is often inapplicable, either because the solids are so fine that they immediately choke up any filter-cloth placed in the basket to catch them, or on the other hand they may not be retained at all on any practicable filtering medium. Such problems may be met by the provision of a basket without perforations, as shown in Fig. 21, and this method has close analogies with the employment of a settling tank. Such machines are known as centrifuges, separators, or subsiders. The latter name is employed when the machine is applied to the removal of solid matter only, but we shall use the term " separator " for all purposes.

The action may be best understood by comparison with that of a continuous settling tank. The liquid to be clarified is fed in at one end, deposits its solids and, if the tank proportions and rate of feed are right, the overflow is clear and bright. If a mixture of two liquids is to be handled, a sill must be arranged as in Fig. 22. The heavy liquid sinks and passes below the sill, and the light liquor floats and is caught on the near side of the sill. Its overflow can be somewhat higher than that for the heavy liquor, according to the relative specific gravities, and by arranging the outlets in this manner control can be exercised

74

on the purity of the discharge.   This point will be
dealt with later.

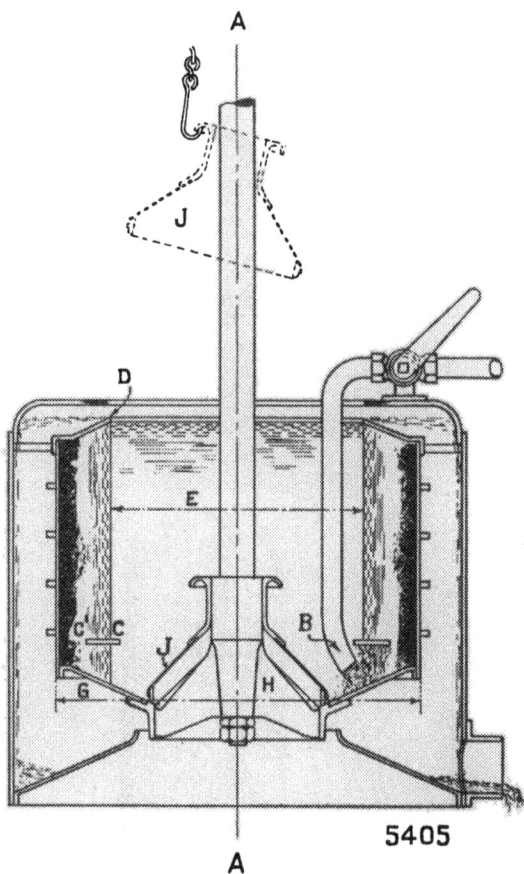

FIG. 21.—Centrifugal Separator—Bulk Type.

Heavy dirt settles to the bottom of such a tank.
Lighter dirt will float on the heavy or the light

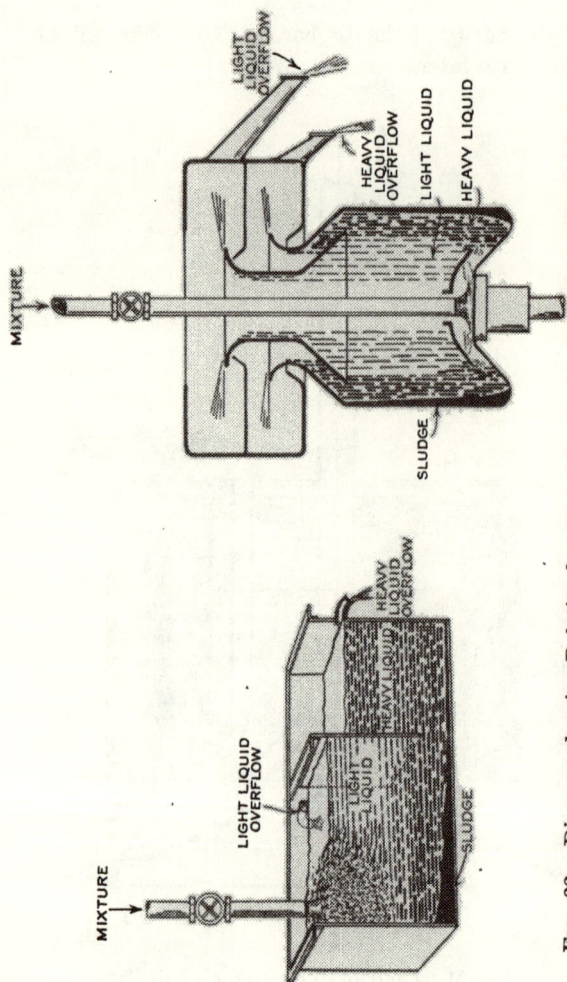

FIG. 22.—Diagram showing Relation between a Separating Tank and a Simple Centrifugal Separator.

liquor according to its specific gravity. In the former case, it will build up on the near side of the sill until it forms a layer deep enough to reach to the sill bottom and pass away with the heavy liquor. In the latter case, it will flow away with the light liquor. If the mixture is simply liquor and light dirt the latter can be trapped on the near side of the sill, and can be removed from time to time if the light liquor outlet is closed. Such dirt would clearly be more trouble to recover than heavier particles.

Matters proceed in the centrifugal separator in a precisely analogous manner. Gravity effects are not perceptible in comparison with those exerted by the very much greater centrifugal force, and the surface of the liquor is practically parallel with the axis of rotation. The heavy solids go to the sides of the basket, the heavy liquor passes beneath the separating ring to the outer (lower) lip, and the light liquor is retained and discharges from the inner (higher) lip. Light solids are rarely dealt with as the construction of most machines does not lend itself to this purpose, but in the next chapter a special machine is described, in which such floating solids can be handled, if the proportion is not too large.

The flow through such a tank or separator may be conceived as taking place in either of two manners. In the first, the liquor moves bodily from inlet to outlet, while the components settle out. If complete separation has not taken place by the time the sill or diaphragm is reached, a mass of partially settled liquor grows until it overflows. According to the other method, the incoming mixture takes its place according to its specific gravity between the separated or partially separated layers; and the rate of motion of a separating particle must be greater than the velocity of the liquor passing from one layer to another. In point of fact both actions occur, and both considerations lead to the same result.

The rate at which a mixture can be separated in a tank depends on its surface area rather than on its depth, for if its depth is doubled a given particle has to move twice as far in order to separate, and therefore the liquor must take twice the time to pass through, so that the volume per hour is unchanged. In a centrifuge, if the bowl diameter is constant, an increased depth does not correspond to a proportional increase in volume, so that the argument is strengthened. Increasing the depth enables the plant to be run for a longer period before it need be cleaned out, and also allows the rate of feed to be increased temporarily by an accidental fluctuation, since a certain volume of unseparated liquor must be built up before it overflows.

It is clear that rapid settling particles (whether larger or heavier) will separate nearer the feed end of the tank or bowl, since the slower ones are carried further forward before they can drop the required distance. Hence a grading occurs which can often be made use of. It is possible to go further, and so select the rate of feed, or in some cases the centrifugal effect, that certain classes of particle are removed, while others are left in the liquor. This grading is not always perfect, first because particles of different specific gravity may so differ in size that their settling rate is equal, and also since slow and heavy particles may begin their fall from different heights in the stream, and come down at the same spot. It is clear that the layer of incoming liquor should be kept shallow, in order to avoid the latter difficulty, and should also be fed as smoothly as possible into the bulk.

We may now consider how it is that the liquor surface is so nearly parallel to the axis of rotation. The real shape is a parabola. If it is assumed that the axis is vertical, it is obvious that at any point in any horizontal plane the centrifugal pressure in a horizontal direction must be equal to the pressure due to the gravity head. If we select a plane passing through the vertex, we have

$$\frac{\omega^2 r \cdot S_L \pi r^2 dh}{2\pi r \cdot dh} = S_L h g$$

or
$$h = \frac{\omega^2 r^2}{2g} \qquad . \qquad . \quad (1)$$

Here r is the radius of the point taken, h the height of the liquor surface above the vertex. For all ordinary purposes this vertex is found to lie so far below the basket (except at low velocities), that the wall of liquor may be considered as vertical. A similar argument shows that if two or more different liquids are present the shape of the face for each liquid will be the same as if it alone were present. In the rare cases where the need is felt, formula 1 will enable the position of any point on the surface to be found.

We can now pass to a more detailed consideration of the laws governing the rate of settlement of particles in suspension. A particle falling in a vacuum will continue to accelerate all the time, according to the ordinary rule. If it falls in a viscous medium, even in air, it will finally attain a speed at which the frictional resistance equals the accelerating force, and its velocity remains constant. In many cases of subsidence this constant rate is attained almost instantly, and the period of accelerated falling can be neglected.

For a comparatively rapidly falling particle the resistance is almost wholly due to fluid pressure on the projected area of the moving particle, and is usually calculated empirically as on left below :—

Fluid resistance = downward force

or
$$k_f \pi a^2 \rho \frac{v_s^2}{2} = \tfrac{4}{3} \pi a^3 (\sigma - \rho) C g$$

Whence
$$v_s = \sqrt{\frac{8 C g a}{3 k_f} \left(\frac{\sigma - \rho}{\rho} \right)} . \qquad . \quad (2)$$

Here k_f = a factor—usually 0·5.*

a = radius of particle in cms.

v_s = velocity of subsidence in cm. per sec.

σ = density of suspended particle—grms. per cu.-cm.

ρ = density of surrounding liquor—grms. per cu.-cm.

For very slow subsidence, the viscous resistance only need be considered, as then it is far greater than the fluid pressure. This resistance is given by Stokes † as

$$- F_s = 6 \pi \mu' \rho a v$$

* Rankine, " Applied Mechanics," p. 598.

† "Mathematical and Physical Papers," Vol. 3, pp. 59–60.

Where μ' is the kinematic viscosity (sq. cms. per sec.).

On equating to the downward force :—

$$v_s = \frac{2Cg}{9} \cdot \frac{a^2}{\mu'}\left(\frac{\sigma - \rho}{\rho}\right) \qquad . \qquad . \quad (3)$$

This is more generally written

$$v_s = \frac{2Cga^2(\sigma - \rho)}{9\eta} \qquad . \qquad . \quad (3a)$$

Where η is the absolute viscosity in dynes per sq. cm., and equals $\mu'\rho$.

Arnold [*] shows that Stokes's law is subject to certain limitations which we have not space to discuss here. If we follow the reasoning of Allen [†] we obtain the following critical values of a and v_s. There is a transition stage, but if a is less than $0\cdot6\ a_c$ (i.e. $v_s < 0\cdot36v_c$), the velocity will approximate to $(3a)$ above.

$$a_c = \sqrt[3]{\frac{9\eta^2}{2Cg\rho(\sigma - \rho)}} \qquad . \qquad . \quad (4)$$

$$v_c = \sqrt[3]{\frac{2Cg\eta(\sigma - \rho)}{9\rho^2}} \qquad . \qquad . \quad (5)$$

η for water at $15°$ C. is $0\cdot0115$. The following table is calculated from the above equations :—

Critical Sizes and Speeds of Suspended Particles.

Centrifugal effect C.	$\sigma = 2$ grams per cu. cm.		$\sigma = 0\cdot9$ or $1\cdot1$ grams per cu. cm.	
	Critical radius a_c mm.	Critical velocity v_c cms. per sec.	Critical radius a_c mm.	Critical velocity v_c cms. per sec.
1	0·085	1·4	0·180	0·6
500	0·011	10·8	0·022	5·0
5,000	0·005	23·2	0·011	10·8
50,000	0·002	50·0	0·005	23·2

* *Phil. Mag.*, 1911, p. 755.
† *Ibid.*, 1900, p. 324, " Motion of a Sphere in a Viscous Liquid."

Since the rate of flow through a subsider depends on the smallest and slowest moving particles to be removed, it would appear that in most cases Stokes's Law will hold, and the speed of separation will be proportional to the centrifugal effect.

This condition of slow free falling, in a viscous medium is not, however, the whole of the story. After a time the particles approach so closely that they impede each other, and a new rate is set up. Walker, Lewis and McAdams give details of certain investigations by Rollason.* The settling period is divided into three stages—free settling, a transition period, and impeded settling. The points at which these begin are defined by the ratio $\frac{h}{h_0}$ reaching certain fixed values, at the first of which the transition period begins, and at the second the impeded period. h is the height of the top layer of the settling particles, h_0 is the height to which they sink after an indefinitely long period, and defines the actual amount of suspended particles which are present per unit area.

The time necessary to drop between two fixed values of $\frac{h}{h_0}$ was found to be proportional to h_0 in all three stages, thus we still keep to the original relation—the time for settlement is proportional to the depth of the layer, and the area retains its old importance.

Other tests made recently by the writer showed that some suspensions may not always obey this

* " Principles of Chemical Engineering," pp. 331, 332.

rule, especiålly if they consist of particles of various sizes. The suspensions in question were of calcium carbonate. That of 20 grams per litre had a very definite free falling period, lasting for a long time, until the upper surface of separation merged into a lower one marking the limit of the larger particles. A suspension of twice the strength had no free falling period, but entered at once into the impeded stage. There was some disturbance due to minute air bubbles which were very difficult to eliminate. It would appear therefore that the type of settling, whether free or impeded, may depend somewhat on the initial conditions if different concentrations are considered, and not merely on $\frac{h}{h_0}$. Reliable experimental work is much needed and may provide some good outlets for the energies of our various schools of Chemical Engineering. In the meantime we will assume that for a similar initial concentration of the same suspension, the average rate of fall to a given final concentration is the same, so that the time of settlement is proportioned to the depth, and flow to the settling area.

So far we have assumed that the only force opposing settlement is that of the liquid displaced. Perrin has shown that Stokes's law holds for particles falling in a cloud, even though Brownian motion is present. He has also shown that particles in suspension exert a mutual pressure in accordance with the gas laws, and thus behave like molecules in solution, with the difference due to their size. If such particles are extremely small, and concentrations great, this pressure may have to be taken into account.

According to his reasoning,* the change in concentration due to a small difference in depth is

$$\frac{K_0}{K_h} = 1 - \frac{N_A}{RT} m \frac{\sigma - \rho}{\sigma} gh \quad . \quad . \quad (6)$$

Where K_0 = No. of particles per cu. cm. at datum.

K_h = No. of particles per cu. cm. at depth h cms. above datum.

N_A = Avogadro's number = 62×10^{22}.

m = mass of particle (grams).

R = gas constant ($= 83 \cdot 2 \times 10^6$).

T = absolute temperature. °C.

Perrin shows that (6) may be transformed into (7) below, and is then valid for any value of h. The following application to the centrifuging of colloidal suspensions is that of Ayres.† Its exactness fails for high concentrations, and the numerical results must be regarded as pointers only.

$$\text{Log}_e \frac{K_0}{K_h} = m \frac{(\sigma - \rho)}{\sigma} gkh \quad . \quad . \quad (7)$$

Where $\quad k = \frac{N_A}{RT} = 3 \times 10^{13}$ approx.

If we write $\quad \alpha = m \left(\frac{\sigma - \rho}{\sigma} \right) gk$

Then

$$K_h = K_0 e^{-ah} \quad . \quad . \quad . \quad (8)$$

If K_i were the initial concentration, and h' a fixed overall depth, then—

$$h' K_i = \int_0^{h'} K_h dh = \frac{K_0(e^{-\alpha h'} - 1)}{- \alpha}.$$

* "Atoms," p. 93.

† American Institute of Chemical Engineers, *Transactions* 9, pp. 203 *et seq.*

Hence $K_0 = h' K_i \left(\dfrac{\alpha}{1 - e^{-\alpha h}} \right)$. $\qquad$. (9)

and $\qquad K_h = h' K_i \left(\dfrac{\alpha}{e^{\alpha h} - e^{\alpha(h - h')}} \right)$ $\quad$. (10)

Now $\qquad m = \dfrac{4}{3} \pi a^3 \sigma$

Therefore $\alpha = \dfrac{4}{3} \pi a^3 (\sigma - \rho) g k$

$\qquad\qquad = 10^{17} a^3 (\sigma - \rho)$ approx. $\quad$. (11)

α is readily found for any particular case, and the equations hold when the concentrations are expressed as volume percentages. Ayres gives the following values for the top, middle and bottom layers, when $h' = 1$, $K_i = 1\%$, and α varies :—

Percentage Concentrations—Gravity.

a.	K_1.	$K_{0\cdot5}$.	K_0.
1	0·58	1·04	1·60
10	0·001	0·10	10·00
100	10^{-40}	10^{-20}	100·00

It will be seen that the condition $\alpha = 100$ represents practically complete settlement. In a centrifuge, at a reference diameter D_0, corresponding to K_0 and C_0, we get (cf. equation 5, p. 12) :—

$\qquad \mathrm{Log}_e \dfrac{K_0}{K_h} = m \left(\dfrac{\sigma - \rho}{\sigma} \right) C_0 g \left(\dfrac{D_0 - 2h}{D_0} \right) kh \qquad (12)$

Putting $\qquad \beta = m \left(\dfrac{\sigma - \rho}{\rho} \right) C_0 g k$

Then $\qquad K_h = K_0 e^{-\beta h (1 - 2h/D_0)}$ $\qquad$. (12a)

and $\qquad h' K_i = \displaystyle\int_0^{h'} K_0 e^{-\beta h (1 - 2h/D_0)} dh$ $\quad$. (12b)

Ayres gives the following approximate solutions for $D = 4$, $h' = 1$.

$$K_0 = K_i \left(\frac{6}{1 + 4e^{-\cdot 5\beta(1-1/D_0)} + e^{-\beta(1-2/D_0)}} \right) \quad . \quad (13)$$

$$K_h = K_i \left(\frac{6e^{\beta h(2h/D_0-1)}}{1 + 4e^{-\cdot 5\beta(1-1/D_0)} + e^{-\beta(1-2/D_0)}} \right) \quad . \quad (14)$$

If the highest available $C_0 = 40,000$

Then $\beta = 4 \times 10^{21} a^3(\sigma - \rho)$.

For a 1% dispersoid treated in a centrifuge having $D_0 = 4$ cm., and $h = 1$ cm., Ayres gives the following table :—

Percentage Concentrations—Centrifuge.

β.	K_1.	$K_{3\cdot 5}$.	K_0.
1	0·80	0·95	1·36
10	0·06	0·08	6·00
100	10^{-20}	10^{-10}	6·00

It will be seen that the results are analogous to gravity subsidence, and the condition $\beta = 10$ corresponds to very fair separation, while $\beta = 100$ gives a practically perfect separation.

The time factor has been investigated by Ayres, and those interested in his reasoning should refer to the original paper.

For a centrifuge of diameter 4 cm., and a layer 1 cm. thick, he obtains the time in seconds as :—

$$t = \frac{1 - \varphi\beta}{\theta(\beta) \left[1 - \frac{\varphi(\beta)}{2} - (1 - \varphi(\beta)) \frac{e^{\beta\phi(\beta)\left(\frac{\phi\beta}{2}-1\right)}}{\beta\varphi(\beta)} \right]} \quad (15)$$

Here $\quad \varphi(\beta) = \dfrac{1 + 4e^{-\cdot 375\beta} + e^{-\cdot 5\beta}}{6}$.

$$\theta(\beta) = K_i Cg \frac{2a^2(\sigma - \rho)}{9\eta}$$

In obtaining $\theta(\beta)$ Ayres makes K_i a weight percentage. In many respects his reasoning is obscure, but the results are said to agree with experiment, and are summarised below :—

With a centrifugal effect of 40,000, a viscosity of 1, $\sigma = 2$; $\rho = 1$, it appears that a 1% suspension of particles 1μ in radius would take 12 seconds to settle; while if the radius were only $0 \cdot 1\mu$ 20 minutes would be required.

If the medium were only as viscous as water the times would be $0 \cdot 1$ second and 11 seconds, respectively. Moreover a particle of radius $10\mu\mu$ would take 18 minutes, but one of radius $\mu\mu$ would require 33 hours. (Note : $\mu = 10^{-3}$ mm.)

It is clear that in dealing with such colloidal suspensions and dispersions, very high centrifugal force is in theory capable of giving results which differ in kind from those obtainable by smaller degrees of force. Not only are separations obtained more rapidly, but if the particles are exceedingly minute, a more effective settlement is given.

In many cases the work of the centrifugal separator has to be made easier by pretreatment of the liquor to be handled. When very fine particles are in suspension, the amount of interfacial surface is enormously increased, and any reduction of the surface tension weakens the forces which would otherwise cause the particles to coalesce. It is also supposed that stabilising substances form protective films at the surface of the particles by adsorption, or may cause them to become electrically charged, and so repel one another. Many substances, such as soaps, gums, gelatine, albumen, etc., have the effect of hindering the settlement of small particles, even if present in very small amounts. The presence of finely divided solids such as carbon, which are unequally wetted by the liquid components of a mixture, may tend to stabilise them in the form of a difficult emulsion.

While many suspensions and emulsions will break up under centrifugal force, others must be dealt with in such a way as to remove the stabilising agent, or destroy its effect. Coagulants, such as acids, alkalis, or electrolytes may be employed, while in some cases a partial concentration followed by dilution will permit the emulsifying agent to be washed out to such effect that a second centrifugal concentration will be complete.

Two types of emulsion are possible with any pair of liquids such as oil and water. We get an oil in water emulsion with a water-soluble stabilising agent, and water in oil with an oil-soluble agent. Either emulsion can sometimes be effectively broken by the addition of an agent tending to change the type.

Agitation, or vibration, according to its amount or amplitude, duration and frequency, can either make or break an emulsion. Heat or electrical treatment also find application for breaking up difficult products.

The difference in density between the particles can sometimes be increased by dissolving salt in one component. Reducing viscosity by heating also permits the particles to separate more rapidly.

The whole question of dealing with these matters

is a very large one, and the reader who is interested will do well to refer to the voluminous literature on this subject.

We have now discussed at length the behaviour of the suspended particles, and can turn to one or two points in connection with the machine.

One important matter is the relation of the diameters of the light and heavy liquor overflows to the diameter of the separating plate.

Let D_S be the diameter of the separating plate, D_H and D_L those of the heavy and light liquor overflows. The obvious condition is that the mass of a ring of liquor of unit width on the heavy side shall be equal to that of a similar ring on the light side; since the whole of the mass of each ring may be treated as being concentrated at the diameter of the lip.

Then $\quad \frac{\pi}{4}(D_S{}^2 - D_L{}^2)\rho_L = \frac{\pi}{4}(D_S{}^2 - D_H{}^2)\rho_H.$

Here $\rho_L \cdot \rho_H$ are the densities of light and heavy liquors respectively. D_S and D_L are usually fixed by general considerations of design. We then have

$$D_H = \sqrt{D_S{}^2 - (D_S{}^2 - D_L{}^2)\frac{\rho_L}{\rho_H}} \quad . \quad (16)$$

If we look into this relation, we find that the smaller we make D_L the greater difference we get between D_L and D_H, especially if D_L be quite near the centre. Consequently it is good to arrange the discharges at as small a radius as possible, especially if the difference in specific gravities is slight.

It is obvious that the formula gives the extreme limit of difference which will still permit light liquor to flow. In theory the result of any overfeeding would be a full volume discharge over the heavy discharge lip, and none over the light discharge lip, as the full column of pure light component must be supported by a pure column of heavy component.

Something a little less than this is usually desirable, and in fact the separator will work on any difference in diameter less than the calculated one. There will then be a certain excess of heavy liquor or mixture on the light side of the lip, and the light liquor will stay in the bowl a shorter period. There is more allowance for temporary variation in the rate of feed, if this is about the maximum ; but if overfeeding goes on, then mixture or only partially separated liquor may pass away from either discharge. It is clear that various disturbances may arise in a continuous separator which do not appear in a tank or a small scale test-tube type machine. There may be cross currents, short circuiting, irregular motion or surging, which must be guarded against.

No recent work seems to have been published as to the relations between the various factors in actual machines, and we are limited to those dealing with milk separation.

Richmond * gives the following relation for the output of a milk separator :—

$$f = k_1 \times k_2^{(40-t)^{38}_t} \times k_3^{\text{F}} \times \frac{\sqrt{Q^3}}{N^2} \quad . \quad (17)$$

* " Dairy Chemistry " (1920), p. 410.

Where f = % fat in skim.

    k_1 = a constant = 8155 for one separator.

    k_2 = 1·035 to 1·05. k_3 = 1 to 1·05.

    t = temperature (° C.).

    F = % fat in cream.

    Q = flow (gallons per hour).

This is on the general lines of work previously done by Fleischmann.* It appears that instead of $\sqrt{Q^3}$, some observations on other types of separator give Q, or even $\sqrt{Q}$.

The factor k_1 appears to depend on the size of the drum, and thickness of layer. It is also affected by the specific gravities of the serum and the milk, and on the units. k_2 depends on the viscosity of the milk serum, and on the friction in the drum. k_3 is appreciable where adjustment is made at cream outlet. It depends on the viscosity of the cream, and on the outlet friction.

It is interesting to note the following specific gravities given by the authorities.

Skim milk at 15° C. = 1·018 s.g.

          20° C. = 1·019 s.g.

          25° C. = 1·0197 s.g.

Fat at   15° C. = 0·92 s.g.

Fat at boiling point = 0·865—0·868 s.g.

Clearly a rise in temperature not only reduces viscosity, but increases the difference in specific gravity. Above 25° C., however, the effect increases more slowly.

The following formula relating the amount of fat in the cream and skim respectively, and the

* " Book of the Dairy," p. 143.

amount of cream taken from a milk of given fat content, is of interest.   The derivation is not given, but the reader may easily prove it for himself.

$$F = \frac{100(f_i - f)}{R} + f \qquad . \qquad . \quad (18)$$

Here f_i = % initial fat in milk.
R = % cream recovered.

Finally we may remark on the power absorbed by the liquor escaping from the bowl.   The discharge lips should be as close to the axis as possible, in order to reduce the velocity of discharge to a minimum.

Let the speed of the escaping liquid be V_0 in ft. per sec., then the horse-power taken is

$$HP = \frac{Q\rho V_0^2}{396000g} \qquad . \qquad . \qquad . \quad (19)$$

CHAPTER V

THERE are on the market a number of different types of these machines and the choice depends very largely on the class of work to be done. They may be divided into :—

> Laboratory Separators.
> Bulk Separators.
> Plate Separators.
> Super-speed Separators.

Small separators of any type may be used for laboratory purposes; but the one most usually employed (Fig. 23) has a head which carries two or more tubes having removable glass interiors of about the size of a small test-tube. These tubes are hinged, and are vertical when not rotated, but fly out horizontally when brought up to speed. The smaller sizes can be clamped to a table, and are operated by hand. The larger sizes are generally run by electric motor.

The glass tubes are usually graduated, and are extremely useful for making tests as to relative amounts of solids and liquor, or of two liquors.

Bulk separators are machines which have a comparatively high settling area and volumetric capacity, and are usually an adaptation of a standard dryer (cf. Fig. 21). Sometimes the only difference is that a plain bowl is substituted for a perforated basket. Such a machine is useful for speeding up settlements which would take place by gravity without too much difficulty. A more definite separation is obtained, in less space, and the solids are ordinarily a good deal dryer than they would be unless a very prolonged gravity settlement were given.

They are quite commonly in use for handling starch. In one instance a 48-in. basket 16 in. deep held about 1 cwt. of starch, and the settled product contained 50% of water approximately. The maximum centrifugal effect was about 600, and settling took place on a 5½-in. wall in about 10 minutes. The settled liquor was not quite clear, and was returned to the process. In this case the starch had been previously purified by passing through settling tanks. Sometimes a centrifugal is used for this purpose, and then the pure starch and the impurer portions settle into distinct layers according to their specific gravity. These can be separated after removal from the machine.

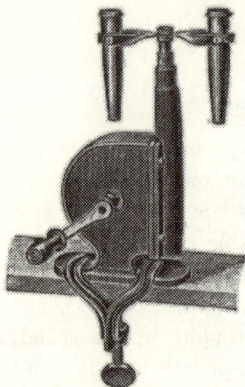

Fig. 23.—Hand Centrifuge. Giving from 2,000 to 3,000 revolutions a minute.

Wool scourings are sometimes given a pre-treatment in these machines, on account of their great capacity for holding dirt, which would otherwise rapidly choke the small centrifugals of greater separating power in which the final separation is accomplished. Here the liquor is run in continuously, and escapes over the basket lip, leaving the dirt behind. This must be removed every two to four hours, according to its amount. The dirt from Australian wools is sandy and readily separated. That from Cape wools is slimier, and also the quantity is greater. The capacity of such a machine is about 15 to 30 gallons per hour per sq. ft. settling area.

The basket is fitted with draining arrangements, in order to run off the separated liquor. A refinement which is useful in some cases is a skimming pipe, which must be adjustable for depth of cut. This skims off the surface of the liquor, and can be employed to separate one component from another. A separating plate may be fitted, if it is desired to separate two liquids; the heavier will pass beneath it, and over the lip into the pan. The lighter will rise to the surface and will be caught by the skimmer. It is possible to work continuously, but often it may be found best to skim at intervals. In any event time should be allowed for a sufficiently deep layer of separated liquor to accumulate before the skimmer is adjusted to its working position.

A certain amount of grease can be recovered from wool scourings by this means, but it is in the form of an emulsion, which separates on standing and cooling. Everything depends, of course, on the liquor being dealt with, and the use of acid or other means of de-emulsification mentioned in the previous chapter will often speed up separation, and make it more complete.

Cross currents should be guarded against when the machine is fed continuously and it is desirable to fit a feed plate at the bottom of the basket, with vanes to bring the entering liquor up to basket speed.

An interesting variation of this type of machine, used for cleaning small quantities of lubricating oil

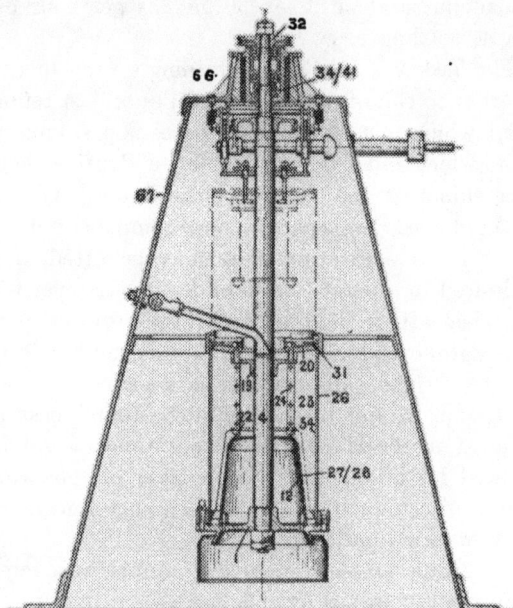

FIG. 24.—Gee Centrifuge.

recovered from swarf, has a plain basket into which a measured amount of water is first placed. This forms a vertical wall inside the basket when speed is attained, with enough space between it and the basket lip to allow a layer of oil to be retained. As the dirty oil is fed on to the bottom of the basket, it is broken up fairly finely and sprays into the water, where the dirt separates and is trapped before the drops rise again to the surface.

Such a machine, with a basket 14 in. in diameter, running at 2,000 r.p.m. will treat some 10 or 12 gallons of oil per hour.

A development of the bulk machine is the Gee Centrifugal Separator (Fig. 24). Its chief characteristic is the use of a very deep basket. The standard size is 3 ft. 0 in. diameter by 4 ft. 6 in. deep, and the speed is 1,000 r.p.m. This increased depth gives a longer settling period, and it is possible to separate the more rapid settling particles from those which come down more slowly. For this purpose the basket has a removable inner lining, in sections, which can be drawn up in a cage out of the basket for purposes of emptying. The coarse portions cling to the top of these sections, while the finer portions settle more slowly and are found near the bottom. It is thus possible to cut the separated material across before removal, and to divide it into various qualities, by which means its value is often much enhanced.

Twenty h.p. is usually provided for accelerating the machine, but 10 to 11 h.p. usually suffices for running purposes when 1,000 to 1,200 gallons are passing through per hour.

In treating clay slip at 1,000 gallons per hour, acceleration in a typical case took 2 minutes, feeding 15 minutes, emptying and restarting 7 minutes, making 24 minutes per cycle. The deposit would work out at about 680 lb. per operation, and its average moisture would be 25%

4

to 30%. This would vary considerably as between the finer and coarser portions.

In handling starch at 1,200 gallons per hour, acceleration took $3\frac{1}{2}$ minutes, feeding 56 minutes, emptying and restarting 14 minutes, a cycle of $73\frac{1}{2}$ minutes. The deposited solids were about 600 lb., roughly 45% moisture. The starch in the overflow was about 0·05%.

One man per machine will usually suffice both for filling and cleaning, but if the cycle is a short one, and there is only a single unit, a boy may be needed as well.

The Hoyle Continuous Separator is designed for the purpose of handling coal washery effluent, which is heavily charged with fine solids, both coal and shale. It consists of a conical outer imperforate drum having its largest diameter at the base, which is fitted with a diaphragm, over the inner rim of which the liquid escapes after treatment. The mixture is fed through a tube into a rotor which projects it all to the sides of the revolving vessel, the solids finding their way to the outer periphery, where they come in contact with the revolving worm travelling at a greater speed than the cone drum.

The basket is approximately 9 in. diameter at the top, and 25 in. diameter at the bottom, and is $23\frac{1}{2}$ in. deep. When operating on washery effluent the speed is 850 r.p.m. and 50 gallons per hour are said to be treated. An effluent containing 41% of solids left with only 1·6%, and the ash content of the recovered solids was slightly less than those on the untreated slurry.

Machines with horizontal spindles have been employed, but have not much to recommend them.

All large diameter machines, however, are limited as to the centrifugal force which they can give by considerations which we have gone into

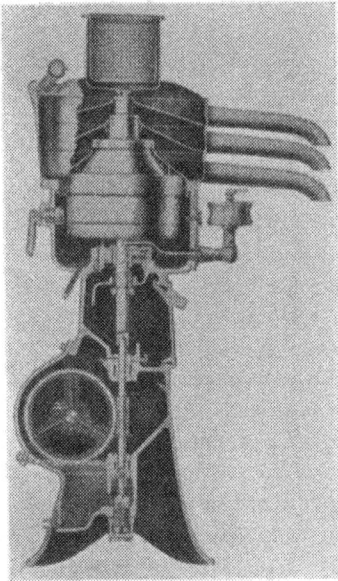

FIG. 25.—Plate Separator.

previously, although on account of the evenness of the loading they can often be run at higher speeds than would be safe if they were to be employed as dryers. The effluent is rarely as clear as that from an efficient filter-press, and the chief advantage is the great capacity for holding deposited solids, and general convenience in operation.

The need for handling difficult separations has led to the development of a class of machine of which the diameter is rarely much greater than 1 ft. (Fig. 25). Such machines can be run at much higher speeds, give high centrifugal effects, and may be referred to as plate separators (Fig. 26). Their development has been largely due to De Laval. The type has become standard for cream separation, but the system is employed very widely.

Clarification is usually as good if not better than that obtainable in a filter-press, and liquors can be dealt with which could not be treated in a press. The capacity for holding solids is on the other hand extremely limited. A typical use is the separation of two liquors of varying specific gravity, with the simultaneous removal of a very small amount of fine slime. The surface area in which separation occurs is greatly increased by filling the interior with plates (Fig. 26), pressed out of thin metal sheets, and provided with suitable spacing pieces to keep them slightly apart, the distance depending on the viscosity of the liquor to be treated. These plates also serve an important purpose in preventing disturbances in the liquor, and in bringing it rapidly up to speed. The feed goes down a central channel, and is distributed upwards through the bowls by suitable passages. A particle has only to move the horizontal distance between the plates in order to separate completely, so that on the underside of a bowl there is a layer of heavy liquid (or sludge) moving outwards (Fig. 27), and on the upper side a layer of light liquid moving inwards. The liquors

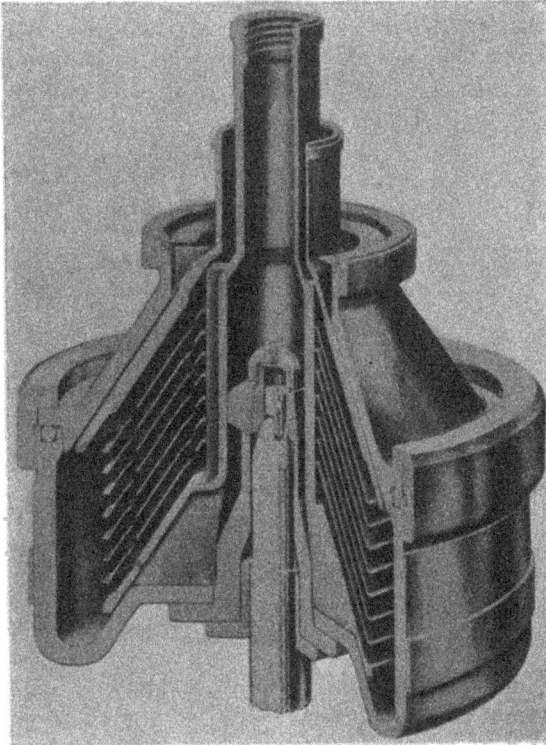

FIG. 26.—Bowl of Plate Separator.

discharge by their respective lips to separate outer covers and spouts. A third cover is often provided to catch any overflow from the feed inlet should overfeeding occur.

Any heavy solid matter which separates is caught in the sludge space between the shell and the outer edges of the plates. It is, however, quite likely to choke between the plates, and in any case the sludge capacity is small. Entering liquors must therefore be as clean as possible, and free from any coarse matter. In the case of milk,

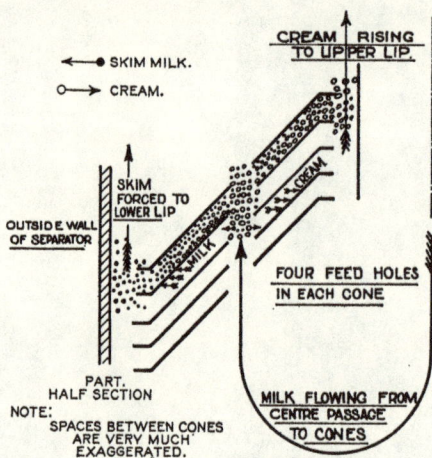

FIG. 27.—Action of Cones in Cream Separator.
(Part : half section.)

if prior cleaning is required, a small cleaning separator is used, very similar in appearance to the final one, but not fitted with internal plates, and having only one discharge lip. The interior is practically plain, except for a distributing plate, and some vanes.

There are some interesting features common to most machines of this class. Some freedom has to be given to the basket, for owing to the very high degree of force exerted, a small eccentricity would cause very high stresses. Sometimes the basket has a little freedom, where it is attached to the spindle. The latter is given a little play by means of an upper spring bearing and a lower spherical joint. The drive is usually through a helical gear.

Various ingenious devices are in use for the purpose of the rapid removal and cleaning of the plates. One system permits them to be removed *en masse* and strung on a wire, when they can be separated and cleaned without losing their relative position. In other cases, however, it is sufficient to clean any deposit from the sludge space, and allow the incoming feed to clear away any deposit between the plates.

In any case stops must be fitted so that it is impossible to replace the bowls in the wrong position, which might mean that ports would not correspond, or balancing be correct.

The results obtainable are dependent on many factors. There are, of course, the size of the machine, and its speed and freedom from fluctuation. Reasonable absence from vibration is also of importance. Then there is the ease with which separation takes place, and substances such as milk, which are liable to deteriorate, should be as fresh as possible when fed, otherwise difficult emulsions may tend to form, and separation will be hindered. The rate of feed should also be as regular as possible.

A Vickcen machine with an 11-in. bowl, running

at 6000–7000 r.p.m. and giving an effect of
5,400 and upwards, will require about 1½ h.p. to
run, and about 2 h.p. for acceleration. Stopping
for cleaning purposes may take as little as 12
minutes, if the sediment is readily removed, and
has collected in the space provided. If the bowls
themselves are choked and dirty, the time necessary
may be much greater.

The following are typical duties :—

Turbine Lubricating oil .	400	g.p.h.
Diesel ,, ,, .	350	,,
Heavy Diesel fuel oil .	350	,,
Used Transformer oil .	200–400	,,
New ,, ,, .	480	,,

Smaller sizes have capacities down to about
1/10th of the above, varying roughly as the square
of the diameter, and as the number of plates,
providing the spacing remains the same. In this
case, of course, cubic capacity and separating
area increase almost in proportion.

Control is provided by a replaceable discharge
lip, and it is often best to make the final adjustment when the machine is installed by trying out lips
of various diameters till the best results are given.

As has been mentioned, machines of this class,
and also the Sharples type described later, are
often used in conjunction with a large diameter
bulk centrifugal, which serves to scour the liquor
before it passes to the plate separator, which otherwise could not be worked economically on account
of choking by solid matter. Two plate separators

may be used in series, one to skim the grease from the effluent in the form of a strong emulsion, and the other to give a water-free grease. About 0·4% to 0·8% of fat is left in the effluent, the recoveries ranging from 25% to 65% on crude wool scourings containing 0·7% to 1·2% of fat.

Machines of this type are also made with internal filters, which operate on the difference of centrifugal head due to the inlet and outlet diameters. Except that the pressure is applied by centrifugal force, they may be classified as pulp filters. Some solids, of course, deposit by settlement. In certain cases centrifugal force will free the filter surface of deposit when it has built up to some extent, but the action is opposed by the liquid pressure and its possibility must be demonstrated by experiment in any given case. The liquid pressure will force the liquor either upwards or downwards as required, or inwards towards the centre of the basket.

Another class of machine is represented by the super-speed separator, or Sharples centrifugal (Fig. 28).

The chief characteristics are the use of extremely high speeds, and the employment of long baskets of small diameter without internal bowls. It is made in several sizes, of which the largest is 4½ in. in diameter by 36 in. long, running at 17,000 r.p.m. (effect 16,950), while the laboratory type, 2 in. in diameter, runs at speeds up to 40,000 r.p.m. (effect about 41,250).

The separating area is much less than in the plate type of machine just described, but this is
4*

compensated for by the large centrifugal force available.

The basket is supported by a tapered spindle, carried on a ball bearing at its upper end. This bearing is of the self-aligning type, and allows the

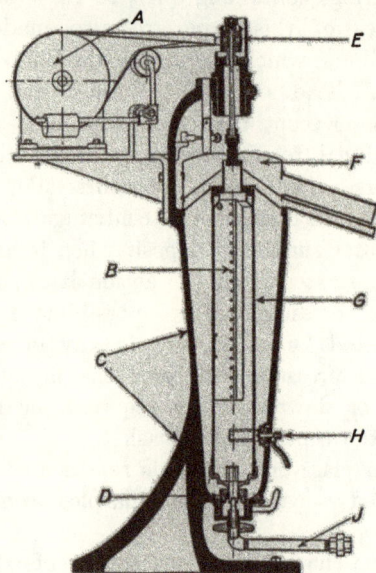

Fig. 28.—Sharples Super-centrifuge.

basket to balance itself within limits when running. The lower end of the basket has a spigot piece through which it is fed by an entering jet. This spigot passes through a frictional control, or steadying bearing at the bottom, the function of which is to damp out any excessive vibration on precession. This control can slide about in its

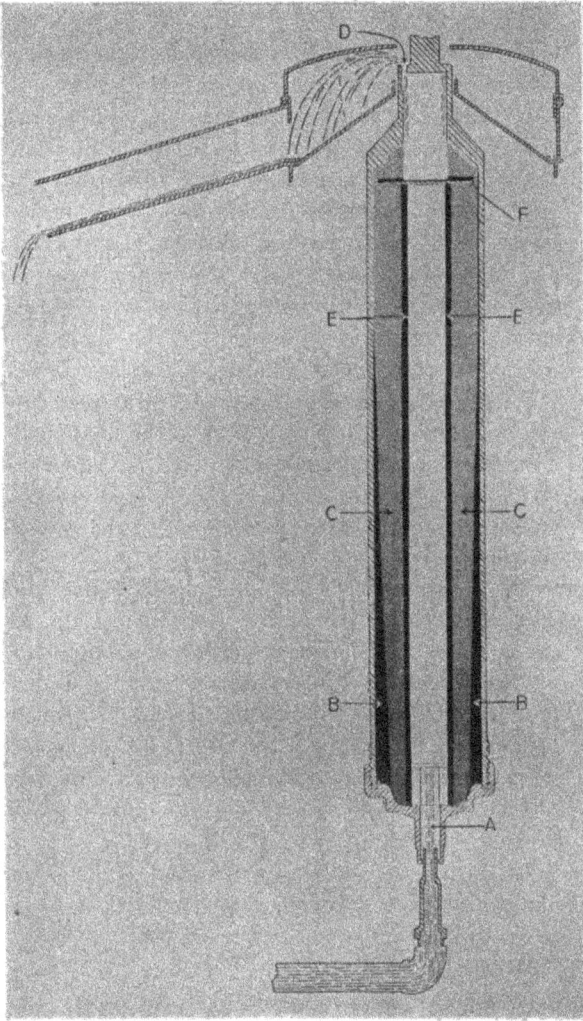

Fig. 29.—Separation of Solids in Super-centrifuge.

A = Feed inlet.   D = Clarified effluent.
B = Heavy solids.   E = Light solids.
C = Liquid.   F = Retaining disc.

housing, but its freedom is limited by friction, controlled by powerful springs. It is made in two types—one is suitable where the feed can act as a lubricant, and the other for forced lubrication.

The entering jet meets a baffle just inside the bowl, which is provided with fins, to bring the liquor up to speed, and to prevent disturbances. The head of the basket has also, when required, light and heavy liquor outlets delivering to separate covers. Control is provided by a renewable separating plate fixed below the discharge head. If a *light* solid has to be retained a retaining plate can be fitted as shown in Fig. 29. The light solids now float on the liquid, and form a cylindrical wall, which breaks up when the bowl is stopped and taken apart. Heavy solids go to the wall in the usual way.

The construction is very ingenious, as the bowl may be rapidly dismantled and opened up for cleaning, for which special brushes are supplied.

. The problem of the best way of handling liquors prior to treatment in this manner has had much attention. For instance, the mucilage from linseed oil requires some seven to eight months for gravity settlement, or two to three weeks if broken by heating to 280° or 300° C. If a $\frac{1}{4}\%$ to $\frac{1}{2}\%$ of fuller's earth is added the mucilage is absorbed and can be removed in the centrifugal, which needs to be cleaned out every two hours or so.

Wool is generally scoured on a series of " bowls " in which it has been found best to transfer the wash liquors periodically from bowl to bowl in counter current to the wool. Thus all the grease becomes concentrated in the first bowl, while the wool is thoroughly rinsed in clean water in the last bowl.

By this means the amount of liquor to be removed per unit of grease is reduced to a minimum. In some cases the water required per lb. of wool washed has been reduced from 10 to 14 lb. to as little as 4 lb. The concentrated liquor is heated to about 160° F., and allowed to stand, or passed through a bulk centrifuge in order to deposit as much mud as possible. It is then passed to a Sharples Centrifugal at about 150 g.p.h. and is separated into soapy water containing about $\frac{1}{4}$% of grease and wet concentrated grease. The latter is washed with hot water in order to free it from emulsifying agents, such as soap, and it is then passed through a second machine, and the grease is recovered free of water (under 0·1%) at a rate of about 50 gallons of finished grease per hour.

Of course, by running at a lower rate, quite fair results can be obtained by only a single passage, but much depends on conditions. At a rate of 125 gallons per hour, recoveries of 15% to 50% are recorded. In one case in which there was only 0·23% of grease in the crude liquor, it was possible to recover 20% of it.

Heavy fuel oil for Diesel engines (boiler oil) can be treated successfully by these machines, even when installed aboard ship. The rate of feed would be from 250 to 350 g.p.h., and the ash content would be reduced from 0·065% to about 0·022%. An overflow for any separated water would be provided.

In the case of water-gas tar, finely divided carbon acts as an emulsifying agent, but its density causes it to separate almost as soon as the liquor enters the basket. The emulsion is then easily separated into oil and water, and about 170 gallons per hour can be dealt with. Gravity separation is inefficient and takes nine to twelve months to complete.

Wax may be recovered from heavy cylinder stock, by running brine into the machine with the oil. The brine forms a layer which isolates the wax from the basket wall, and carries it with it to the heavy liquor discharge. The oil forms a third layer on the inside, and comes away, wax free, from the inner discharge. This oil will now have a cold test as low as 19° to 20° F., as against 50° F. with older methods.

Glue may be clarified by treating the suspended matter with suitable coagulants. They are then found to be of lighter specific gravity than the glue liquor and float, forming an inner layer which is caught by a retaining ring under which the liquor passes to the discharge.

Dry cleaning presents a typical example of the application of the machine to clarification. A pump circulates the cleaning fluid from the washers to the separator, so that a change is effected every three or four minutes. The dirt works out of the garments into the solvent and is retained in the clarifier, from which only clear bright solvent returns to the washer. Washing and rinsing are thus continuous, and take from 20 to 30 minutes. The large machine operates at 600 to 1000 gallons per hour. The necessary store of solvent is much reduced, output is approximately doubled, and the cleaning is excellent, since the dirt is not reworked into the clothes. [The above problem, among others, may also be dealt with by filtration, but it is not within the scope of the present work to discuss the advantages or technique of the procedure.]

Many other examples could be given, but those above will suffice to indicate the general capabilities of the various types of separator. Many problems handled by the super-speed centrifugal are also handled well by the plate type, and *vice versa*, so that judgment must be based on the relative performances and economies for the particular case. In most instances the problems that arise have some feature or circumstance which gives them individuality, and in deciding on the details of plant and treatment regard must be had to all the surrounding circumstances.

CHAPTER VI

THE GENERAL PRINCIPLES OF SELF-BALANCING CENTRIFUGALS

Whirling deflections and critical speeds; effect of various factors—" Suspended " machines—Balancing—Precession, disposing factors and cure—Mathematics of both motions.

THE behaviour of self-balancing centrifugals is worthy of a more extended notice than that given in Chapter III. The characteristic motion of such a machine, when accelerating with an unbalanced load, is shown in Fig. 30. The deflection tends to become very great in the neighbourhood of a certain critical speed. Below this point the deflection increases with the speed, but above it the unbalanced mass is opposed to the deflection, which decreases until the basket finally rotates about its centre of gravity (or approximately so) when high speeds are reached. An examination of the laws of vibrating masses shows that they lag half a period behind the disturbing force, whence once the critical speed is passed, but the principle may perhaps be made clearer by an experiment. If a weight is suspended from the hand by a piece of elastic, and the hand is moved up and down in a regular manner, the critical rate is readily found, and above this the weight will be seen to descend as the hand is raised, and *vice versa*. If the motion is very rapid, but in a true line, the weight will remain steady, just as the centre of gravity of the basket finally comes to rest on the axis of rotation.

Governor arms are not subject to this phenomenon, as they are so hung that this lag cannot

112

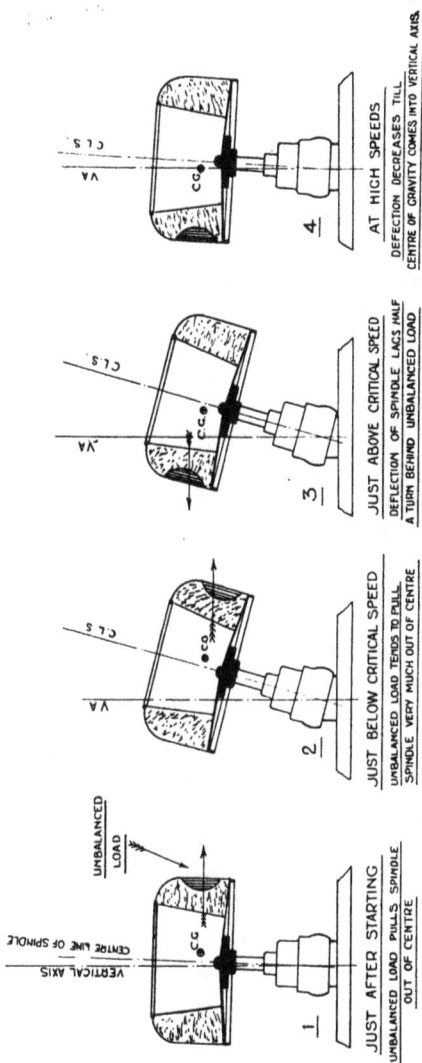

1 JUST AFTER STARTING
UNBALANCED LOAD PULLS SPINDLE OUT OF CENTRE

2 JUST BELOW CRITICAL SPEED
UNBALANCED LOAD TENDS TO PULL SPINDLE VERY MUCH OUT OF CENTRE

3 JUST ABOVE CRITICAL SPEED
DEFLECTION OF SPINDLE LACS HALF A TURN BEHIND UNBALANCED LOAD

4 AT HIGH SPEEDS
DEFLECTION DECREASES TILL CENTRE OF GRAVITY COMES INTO VERTICAL AXIS.

UNBALANCED LOAD

CENTRE LINE OF SPINDLE

VERTICAL AXIS

FIG. 30.—Diagram of Action of Self-balancing Centrifugal, showing Motion of Centre of Gravity and Unbalanced Load with Increasing Speed.

113

occur, and so their deflection increases continually with increase of speed.

If the basket is true, and truly loaded, no whip will appear, but its sensitiveness is mainly dependent on its critical speed. Fig. 31 shows how the deflection curve is affected by increases in this factor. If the critical speed is doubled then the basket must be accelerated through double the number of revolutions per minute in order to cause it to pass through a given range of deflection— *i.e.*, the distance between the ascending and descending parts of the deflection curve is doubled. The final deflection is not zero, but is equal to or nearly equal to the eccentricity, this depending slightly on the basket dimensions.

The critical speed is decreased by decreasing the restoring force of the buffer; increasing the mass of the basket, increasing the spindle length; and increasing the moment of inertia of the basket about a diameter through its centre of gravity, as compared with its moment of inertia about the spindle. These factors are interchangeable to some extent, so that a short spindle is compensated by a weak buffer, within certain practical limits. A stiffish buffer may not be disadvantageous, if this quality be frictional and not elastic. Fig. 32 shows how the theoretical deflection (upper curve) was reduced for varying degrees of friction, as determined by experiment with a model. As friction increases the maximum deflection decreases. Under certain conditions the critical speed may increase, as in the instance illustrated, but the critical period is decreased. Incidentally

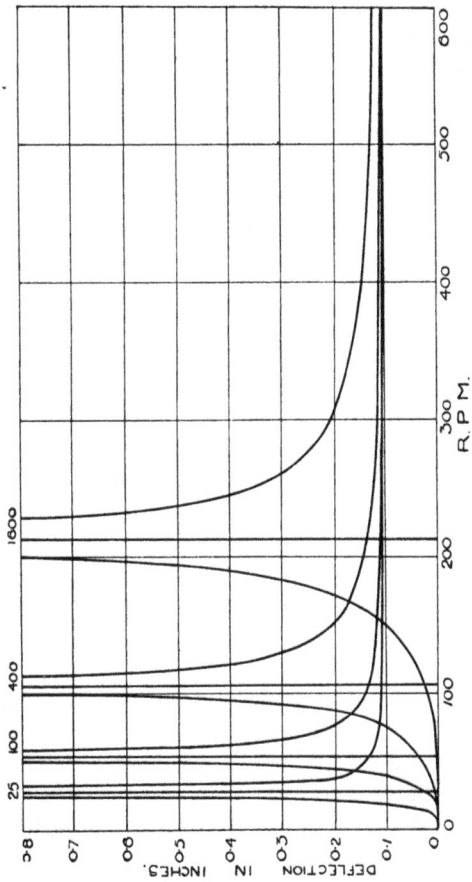

Fig. 31.—Motion of Self-balancing Centrifugal.

Note.—Basket eccentricity is 0·1 inch. Mass 1247 lb. The curves are drawn for restoring forces (from left to right) of 25, 100, 400 and 1600 lb. *per inch* deflection.

the outer point of the basket lags behind the eccentricity under frictional conditions, being 90° behind at the critical speed, and 180° at very high speed. Increased friction causes the position at intermediate speeds to shift a little nearer 90°. This is of importance in balancing, as will be seen.

A heavy basket not only reduces the critical speed, but also the eccentricity for a given out-of-balance loading. Thus a fully loaded basket is less sensitive than a lightly loaded one.

Good acceleration also plays an important part, since it hurries the basket through the critical period before it has time to attain much deflection, unless the eccentricity is very great. This factor is unfortunately limited by the question of power.

Only one critical speed ordinarily appears in the centrifugal, namely, that corresponding to the buffer stiffness, for the shaft should be so stiff in relation to the speed that it cannot set up a secondary vibration. Harmonic critical speeds are therefore absent, but might come into play if a flexible shaft were employed.

Deep baskets are subject to a couple tending to cause them to upset and rotate about their axis of maximum inertia, which is a diameter through the centre of gravity. This is resisted by the stiffness of the shaft, which must be of adequate strength. There is no unusual tendency for such a deep basket to fly outwards, and in fact it centres itself more rapidly than a shallow one. It is, however, subject to large forces if it or its load be balanced defectively, so that deep baskets are not usually employed in fixed-spindle machines,

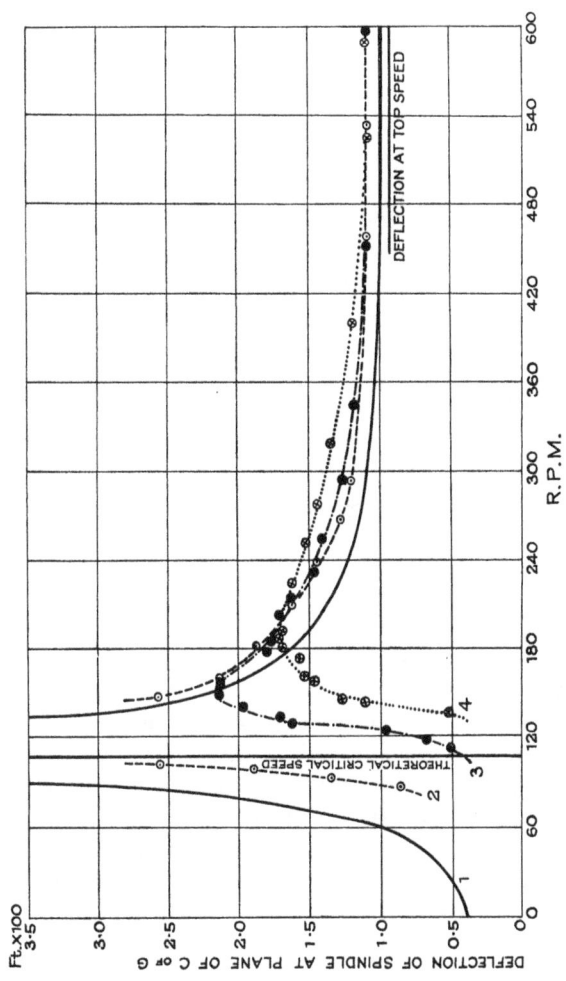

FIG. 32.—Effect of Friction on Whirling. (1) Theoretical : no friction. (2) Experimental with minimum friction. (3) and (4) Increasing degrees of friction.

Note.—The marked rise in the speed of maximum deflection appears to be due to friction in a universal joint used in the experimental machine.

particularly if speeds be high, as then these large stresses would be transmitted to the bearings.

Suspended machines form a separate class. The critical speed is low and depends only on the length of the suspension rods—it is 54 r.p.m. if the rods are 1 ft. long. The mass of the pan is in effect added to the mass of the basket, so that the eccentricity is reduced, and consequently the deflection, so that it will take badly balanced loads. The stress on the bearings is slightly reduced as compared with the ordinary fixed-spindle type, but is not nearly so small as in the self-balancing type. Strong tilting moments can be set up in these machines if their mass centre is too far removed from the plane in which the unbalanced forces act. Strictly speaking, this also applies to the point of attachment of the rods; but since their pull is relatively small, they may usually be located to suit convenience in other directions.

Liquid loads and compensating masses tend to fly to the outer side of the basket, adding to the unbalanced load below the critical speed, and counteracting it above. Unless precautions are taken compensating weights will either hunt or delay action. They are best associated with a good suspension. If several compensating weights are used they set themselves more or less symmetrically opposite to the eccentricity, according to the compensation required.

Good balancing is essential to the best running, as is a true basket. Very fair results (particularly with shallow baskets) are obtained by the ordinary method of adding weights by trial till the basket will not turn when free to rotate on a horizontal spindle, but care should be taken to eliminate friction. Uncompensated couples may still remain (Fig. 33), and a more exact method is to suspend the basket on a spindle of the proper length and find by trial the position and amount of compensation needed to make the basket run true and eliminate critical vibration. The diametral

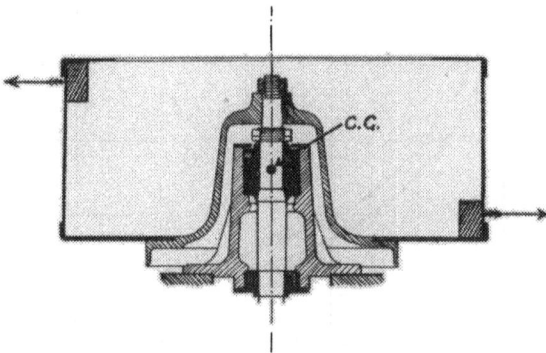

FIG. 33.—Residual Uncompensated Couple.

plane of the unbalanced load can be found by chalk mark at high speed ; or preferably by taking the mean of two positions found by rotating at equal speeds in opposite directions. Kinetic methods can be used for the whole operation, but it is perhaps easiest to balance statically first, and eliminate the residual couples by two equal weights at opposite sides of the basket, of which the one farthest from the buffer must be at the outer side at high speed. Still more accurate and elaborate methods are possible, but are not always as necessary as in turbine work, since the load is usually certain to be somewhat untrue. Simple approximate methods usually suffice, but in any event some experience is needed on the part of the operator when kinetic methods are employed.

Precession takes place when the basket tilts.

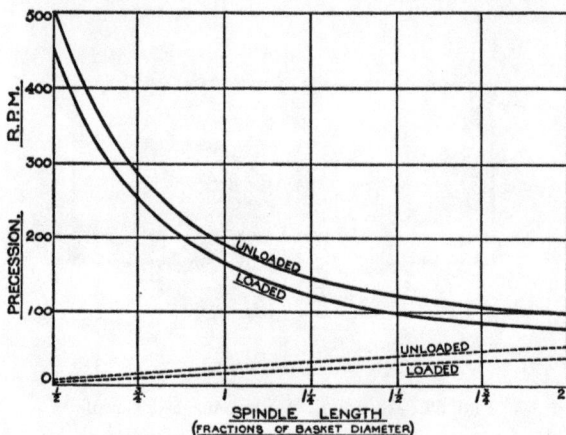

Fig. 34.—Relation between the Precession and Spindle Length for both Unloaded and Loaded Baskets.

See page 123 for the dimensions of the various factors relating to this diagram, and Figs. 35 and 36.

Full lines represent direct precession. Dotted lines represent reverse precession.

It is a comparatively slow gyration of the spindle, about its normal position, which may occur at any speed, though it is most usual at high speeds. It may grow until restrained either by buffer friction or external means, otherwise the basket may strike the outer pan.

The precession usually seen in a centrifugal is in the same direction as the rotation, being the faster of the two possible precessions. It is forced by friction in the transmission or by air swirl. These tend to damp out the slower precession in the reverse direction, which is the one seen in a suspended gyroscopic top. The latter precession may, however, be observed momentarily in a short-

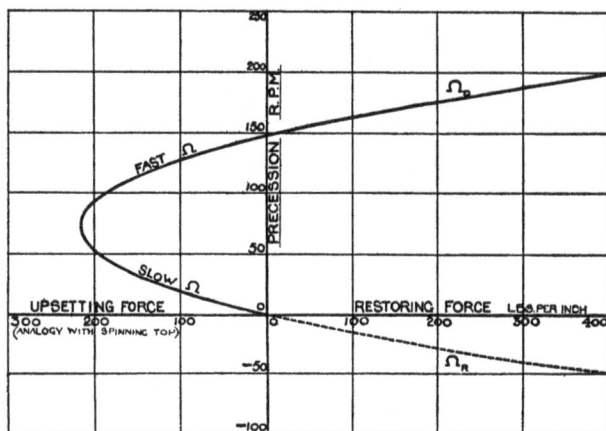

FIG. 35.—Relation between Precession and Force on Basket.

Comparison of self-centring conditions and those of spinning top. Dotted lines represent reverse precession.

spindle machine after a sudden large deflection. The direction of precession does not alter whether the machine be over- or under-driven, since it is always self-centring. It may also be noted that its speed is independent of the deflection, provided that the angle is small.

Air swirl is most noticeable at high speeds, and

with deep baskets. The frictional control of the buffer is weakened by a long spindle, which is most liable to precession for this and other reasons. Since a slow precession requires less energy to start than a fast one, it is helpful to design the machine in such a way that the direct precession is reasonably fast. Here again unduly long spindles are detrimental, as is clearly shown in

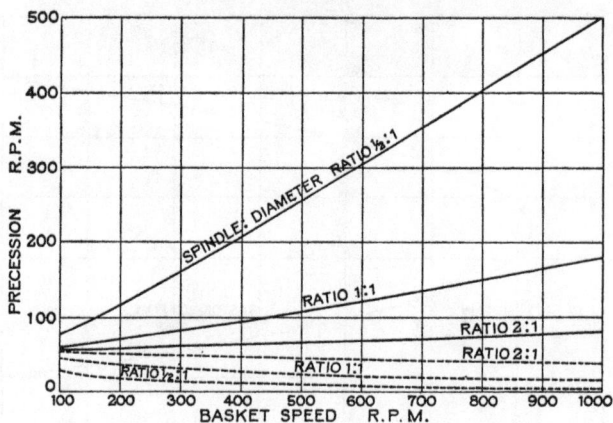

Fig. 36.—Relation of Precession to Basket Speed.
Full lines represent direct precession. Dotted lines represent reverse precession.

Fig. 34. Increased buffer strength helps a little in theory, as is seen in the upper right-hand side of Fig. 35. This diagram also indicates the relation of the two precessions in a centrifugal to those which would occur (left-hand side) if the basket resembled a top, and was subject to an overturning force. In the case of the centrifugal a hindrance to either precession causes it to give way to the resisting force, but in that of a top this only applies to the slower, normal precession.

The effect of speed is shown in Fig. 36. Although precession is faster at high speeds, especially with short spindles, this favourable effect is nullified by the much increased air swirl.

All these curves correspond to the following conditions (see mathematical treatment below).

Diameter = 40 in.

M = 1247 lb. loaded or 687 lb. unloaded.

K_S^* = 17·1 in.   ,,   or 18·4 in.   ,,

K_D^* = 13·7 ,,   ,,   or 14·75 ,,   ,,

R = 1200 lb. per foot deflection (for varying spindle length).

L = 1 diameter (40 in.) for varying buffer strength.

R.P.M. = 900 when other factors vary.

Except in Fig. 34, loaded conditions are always dealt with.

Increasing the basket weight slows the precession, but adds inertia, and may in fact contribute to steadiness.

Experiments are in general agreement with the mathematical treatment. The effects of buffer strength and increased weight of basket were somewhat uncertain, however, and appeared at times to introduce other factors, or variations in friction, which obscure their effect.

Air swirl was found to be diminished by the use of fixed baffles, and by increasing the space between the basket and outer casing.

* See p. 130.

There is, however, no final security against precession unless the external friction is at least equal to any air swirl or friction in transmission. The buffer should provide this, but if precession occurs it may be opposed by laying the hand, or a bar against the spindle so as to hinder its motion. Such a bar is sometimes permanently attached to the machine. Spring buffers alone are unsatisfactory, except, perhaps, for small shallow baskets, as they have not sufficient friction, and if they are used additional friction must be provided in most cases. Rubber, suitably employed, has very fair frictional qualities, and though certain mechanical arrangements may be better in this respect, it is often more convenient and economical.

Pulsations of the correct periodicity, either externally or in the drive, may assist in setting up precession (or whirling), but are rarely troublesome in practice.

It is a well-known fact that a nearly spheroidal top, filled with liquid, will not spin if its depth be even slightly greater than its diameter (unless the difference be such that the depth is at least three times greater). Under such conditions the liquid precession breaks down into turbulent motion, and the top falls. In the centrifugal, any hindrance to precession merely causes the spindle to return to a stable position; the rigidity of the spindle compels the basket and its contents to rotate about it as an axis, and from practical experience it would appear that any unbalanced liquid pressure is not of paramount importance, if speed is high enough to ensure comparatively steady running. Centrifugal baskets of all depths can be run successfully if the general design is correct.

Where tilting is disregarded, we may simply equate the unbalanced centrifugal force to the restoring force, and we get :—*

$$H = \frac{-E}{1 - \frac{Rg}{M\omega^2}} \qquad . \qquad . \quad (1)$$

and by substitution :—

$$H = \frac{-EN^2}{N^2 - N_c^2} \qquad . \qquad . \quad (1a)$$

Also $\qquad N_c = \frac{30}{\pi}\sqrt{\frac{Rg}{M}} \qquad . \qquad . \quad (2)$

Here

H = deflection in feet at c.g.

E = eccentricity of basket c.g. in feet.

R = restoring force of buffer, lb. per foot deflection, at c.g.

g = 32·2 f.p.s.²

M = mass of basket (and load) in lb.

N = r.p.m.

N_c = critical speed (this speed makes H = ∞ in equation 1).

ω = speed :—radians per sec.

The critical period can readily be calculated from the above expressions, and is found to be :—

$$N_{-H} - N_H = N_c\left\{\sqrt{\frac{H}{H-E}} - \sqrt{\frac{H}{H+E}}\right\} \quad . \quad (3)$$

Where N_{-H}; N_H; are speeds for deflection H and $-H$.

* *Trans. Inst. Chem. E.*, 1924, p. 39, etc.   Morley, "Strength of Materials," p. 404.

If we choose $H = 2E$ then the critical range is approximately

$$C_R = 0.6 \, N_c \quad . \quad . \quad . \quad (4)$$

Thus the critical speed defines the critical range.

For a Suspended Machine

$$N_c = \frac{30}{\pi}\sqrt{\frac{g}{L}} \cdot \quad . \quad . \quad (5)$$

Where L = length of suspension rods in feet.

The outer pan swings with the basket and on examining the conditions we find :—

$$H = \frac{-EM}{M + M_P} \times \frac{N^2}{N^2 - N_O{}^2} \quad . \quad . \quad (6)$$

Where

M_P = weight of outer pan.

Equation (6) reduces to (1a) if we make E the eccentricity of the *whole* swinging mass.

The force exerted by the basket on the spindle is

$$M\omega^2 \, (E + H).$$

At very high speeds, from (6), H is very nearly equal to

$$\frac{-EM}{M + M_P} \quad . \quad . \quad . \quad (7)$$

At such speeds, therefore, the force B_S in lb. on the spindle is very nearly

$$B_S = \frac{M\omega^2 E}{g}\left(\frac{M_P}{M + M_P}\right) \quad . \quad . \quad (8)$$

Thus in " suspended " centrifugals the pull on the spindle is reduced for all practical purposes in the ratio of the weight of the outer pan to the total suspended weight, when compared with a bolted-down fixed-spindle machine.

The actual stress on the bearings will of course differ from this, according to the length of the spindle, and the distance apart of its supports.

The deflections and critical speeds, corrected for the tilting of the basket, are more conveniently dealt with after precession has been discussed, since some of the reasoning is the same.

In Fig. 37, the basket is rotating with speed ω about GO, and GO is precessing about YO with speed Ω, in the direction of the arrows, which are to be taken as above the lines they cover. Let the basket have moments of inertia I_S I_B about the spindle GO, and centre of buffer O respectively. If the angle of inclination α is small Ω has a component $\Omega\alpha$ about A′O. Its component about GO is already included in ω. There are thus two angular momentums $I_S\omega$ about GO; $I_B\Omega\alpha$ about A′O.

These can be resolved into components about YO, XO. Only those about XO need be considered since no others are rotated by Ω. The sum of the components about XO is by the ordinary vector rule (cf. " Spinning Tops," Crabtree, p. 86):—

$$I_S\omega\alpha - I_B\Omega\alpha \qquad . \qquad . \quad (9)$$

They are rotated at speed Ω, and therefore the equivalent torque about Z′O is

$$\Omega(I_S\omega\alpha - I_B\Omega\alpha) \qquad . \qquad . \quad (10)$$

This must correspond with the buffer torque, which is positive (clockwise) about ZO. Let B be a modulus, such that Bα gives the buffer torque in foot pounds for a deflection of α radians. Equating this to expression (10) transferred from $Z'O$ to ZO (by change of sign), we get

$$- \Omega(I_S\omega\alpha - I_B\Omega\alpha) = Bg\alpha \quad . \qquad . \quad (11)$$

or $\quad \{I_B\Omega^2 - I_S\omega\Omega - Bg\}\alpha = 0 \quad . \qquad . \quad (12)$

Hence $\alpha = 0$, or

$$\Omega = \frac{I_S\omega \pm \sqrt{I_S^2\omega^2 + 4I_BBg}}{2I_B} \quad . \qquad . \quad (13)$$

Where Ω = precession—radians per sec.

I_S = moment of inertia about spindle—lb. ft.^{2}.

I_B = moment of inertia about buffer centre—lb. ft.^{2}.

B = buffer modulus : torque per radian of deflection, in foot pounds (*including effect of loaded basket on the net buffer pull*).

It should be noted that

$$B = B' \pm ML \quad . \qquad . \quad (14)$$

Where B' = net buffer torque when basket and load are removed.

L = spindle length. Buffer centre to c.g. feet.

The $+$ value is used for top-driven machines and the $-$ value for under-drive.

Also $\qquad B = 12\,Pl^2 \quad . \qquad . \quad (15)$

Where P = force in lb. to deflect loaded basket 1 in. at distance l feet from buffer centre.

It is now possible to proceed to the whirling equations for the case of a basket which tilts. Centrifugal couples are set up which tend to right or upset the basket, according to whether K_S is greater than K_D (*vide* p. 131), or vice versa. The torque thus exerted can be calculated by regarding it as that necessary to maintain a forced precession in which $\Omega = \omega$. Returning to Fig. 37 and equation (10), we now get for the torque about $Z'O$ the value

$$I'_S \omega^2 \alpha - I'_B \omega^2 \alpha$$

or about ZO

$$(I'_B - I'_S)\omega^2\alpha \quad . \quad . \quad (16)$$

Here I'_B, I'_S refer to the balanced portions only of the revolving system.

Now about ZO we also have the buffer torque $Bg\alpha$; and the centrifugal torque due to the out-of-balance weight U, acting in the contrary sense to B. If U be considered concentrated at a radius r, at a distance l from the buffer centre, the centrifugal torque exerted by U, neglecting terms in α^2 is

$$- U\omega^2\{(l^2 - r^2)\alpha + rl\} \quad . \quad . \quad (17)$$

There is also a torque due to the weight of U, also about ZO, and again (in the case of an under-driven machine) opposed to the buffer. This is equal to

$$- Ug(\alpha l + r) \quad . \quad . \quad . \quad (18)$$

The equivalent torque (16) must then be equal to Bga, plus the torques given in expression (17) and (18). Equating to zero we have

$$I'_B \omega^2 \alpha - I'_S \omega^2 \alpha - Bg\alpha + U\omega^2\{l^2 - r^2)\alpha + lr)$$
$$+ Ug(\alpha l + r) = 0$$

Whence

$$\alpha = \frac{- \{U\omega^2 lr + Ugr\}}{I'_B \omega^2 - I'_S \omega^2 - Bg + U\omega^2(l^2 - r^2) + Ugl} \quad (19)$$

Now $I'_B + Ul^2 = I_B$, the moment of inertia of the whole mass of the basket about the buffer centre. Similarly $I'_S + Ur^2 = I_S$. Also $H = L\alpha$. Hence, dividing by ω^2 we have

5

$$H = \frac{-ULr\left(1 + \frac{g}{\omega^2}\right)}{I_B - I_S - \frac{(B - Ul)g}{\omega^2}} \quad . \quad . \quad (20)$$

Hence $N_C = \frac{30}{\pi}\sqrt{\frac{(B - Ul)g}{I_B - I_S}} \quad . \quad . \quad . \quad (21)$

Here H = deflection in feet (at c.g.)

U = out of balance mass lbs.

l = distance of plane of U from buffer centre (ft.).

r = radius of U (feet).

ω = speed-radians per sec.

B = buffer torque. Lbs. per radian deflection.

$I_B I_S$ = moments of inertia about buffer and spindle.

N_C = critical speed. r.p.m.

This is for an under-driven machine. In the case of a top-driven machine equation 18 changes sign, and 21 becomes :

$$N_C = \frac{30}{\pi}\sqrt{\frac{(B + Ul)g}{I_B - I_S}} \quad . \quad . \quad . \quad (21a)$$

The term ul may almost always be neglected in comparison with B, and we can expand I_B and I_S, so that (21A) may be written

$$N_C = \frac{30}{\pi}\sqrt{\frac{Bg}{M(L^2 + K_D^2 - K_S^2)}} \quad . \quad (21b)$$

This applies to both under- and over-driven machines.

Here K_D = radius of gyration of basket and load about a diameter through c.g. (in feet).

K_S = radius of gyration of basket and load about spindle (in feet).

When $K_D = K_S$ equation (21b) gives the same result as (2), and the result is the same as if no tilting occurred, or if the whole mass of the basket were concentrated at a point.

The most useful values for l and r are L and $\frac{D}{2}$, when U is considered to be situated at the end of a diameter through the basket c.g. Further, we can factorise on the lines of equation (1a), when we get

$$H = \left(\frac{-UL^2D}{2(I_B - I_S)}\right)\left(1 + \frac{91\cdot2g}{LN^2}\right)\left(\frac{N^2}{N^2 - N_0^2}\right) . \quad (22)$$

This, again, is for an under-driven machine. For an over-driven machine the central term becomes

$$\left(1 - \frac{91\cdot2g}{LN^2}\right)$$

The amount of whip is now proportional to the high speed deflection given by the first term of (22)

If $K_D = K_S$, (22) reduces to E at high speeds, and the rotation will be about the centre of gravity, just as in the simple case.

At moderate and high speeds the critical speed retains its importance as a governing factor for the range of whirling. At lower speeds the central term has an effect which tends to increase the period of whirl. The critical speed N_{MC} giving a minimum critical period is found by differentiating and criticising in the usual manner, since (22) gives more than one maximum and minimum value. The one in question is found to be

$$N_{MO} = \sqrt{\frac{91\cdot2g}{L}} \qquad . \qquad . \quad (23)$$

This applies to the under-driven type.

In substance N_{MC} represents the critical speed below which the gravity couple due to U suffices to produce a substantial deflection. No such minimum occurs with top-driven machines, under practical conditions, as N_O can never be less than that due to the gravity couple alone.

In comparing the results given by formula (2) and formula (21), it will be found that the simple form is very tolerably accurate for the majority of cases, and will usually be sufficiently correct except for fairly short spindle lengths, or baskets of exceptional shape.

The effect of friction may now be considered. It is usual to assume the drag to be proportional to the velocity, and it is then equal to

$$\mu'g H \omega.$$

Here $\mu' = $ frictional drag in lb. per ft. sec.

In order to maintain steady motion a couple equal to the frictional torque must be applied by the driving mechanism (Fig. 38). This couple may be assumed to have E for its arm, and a pair of forces T, each equal to

$$\frac{\mu'g H^2 \omega}{E}.$$

At the centre of gravity one force T combines with the centrifugal force $m\omega^2 r$ (where r is the radius of whirl of the c.g. in feet) to give a resultant which must be in the line of E, for equilibrium. This, when combined with the other force T, transfers the original centrifugal force to the spindle or basket centre, parallel to its old direction.

The horizontal component must equal the drag, and the vertical component the buffer pull. That is

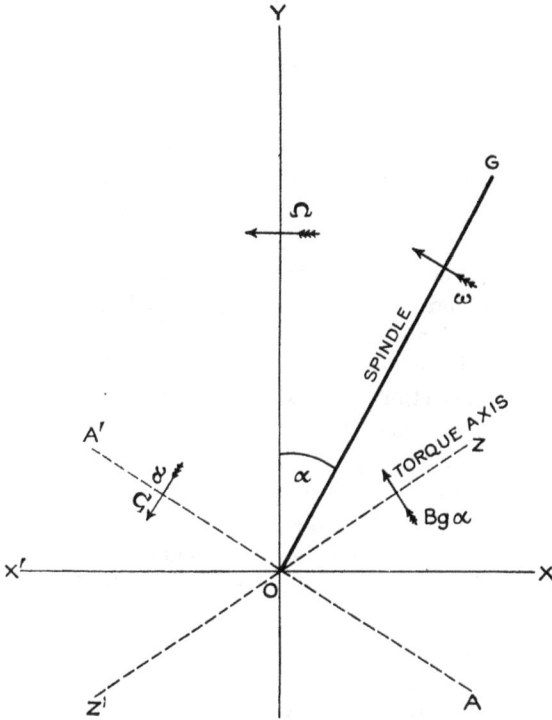

FIG. 37.—Relation between Momentum, Torque, and Precession for an Under-driven Centrifugal.

$$M\omega^2 E \sin \alpha = \mu' g H \omega \quad . \quad . \quad . \quad (24)$$

and $\quad M\omega^2(H + E \cos \alpha) = RgH \quad . \quad . \quad (25)$

Where α = angle of advance in radians.

Evaluating for $\sin \alpha$ and $\cos \alpha$, we obtain

5*

$$\text{Tan } \alpha = \frac{\sin \alpha}{\cos \alpha} = \frac{\mu'g\omega}{Rg - M\omega^2} \quad . \quad . \quad (26)$$

Since $\cos \alpha = \dfrac{1}{\sqrt{1 + \tan^2 \alpha}}$ we can solve equation (25) for H, and obtain

$$H = \frac{EM\omega^2}{\sqrt{(Rg - M\omega^2)^2 + (\mu'g\omega)^2}} \quad . \quad (27)$$

Similar values for tan α and H have been obtained by Frith and Buckingham,* who treated the question as one of vibration.

It will be seen that H is always finite if friction of this kind is present. At high speed H becomes equal to E, and α to 180°. The effect of increasing friction in this case is to raise the speed of maximum deflection very slightly, but the speed at which α becomes 90° is unaltered and corresponds to the natural speed of transverse vibration of the system. In many practical cases they will nearly coincide. Other cases which may be studied on the above lines are " solid " or constant friction, and friction in the line of the buffer pull. These do not present any special difficulties. With constant friction there is no definite deflection until a speed is reached at which the unbalanced centrifugal force has grown sufficiently to overcome it. Friction in the line of the buffer pull, such as may occur in a universal joint, affects the speed at which α becomes 90°, and also the deflection. In all practical cases the critical period and deflection diminish with increasing friction.

* " Vibration in Engineering," pp. 24 and 26. *J. Inst. Electrical Engineers*, 1924, p. 107.

CHAPTER VII

PRINCIPLES OF THE CENTRIFUGAL CLUTCH, ETC.

Inertia and other resistances—Coefficient of friction—
   H.P. of clutch—Mathematical treatment of time
   required for acceleration with clutch, and also with
   Pelton Wheel. Acceleration allowance.

THE uses of the centrifugal clutch were referred
to in Chapter III, but some problems remain which
are worthy of study.

In starting up a centrifugal various resistances
must be overcome. The most important is the
inertia of the basket and load. As a rule a good
deal of liquor is lost during accelerating, and in
some cases the load compacts itself, so that this
factor is somewhat variable. There is the bearing
resistance, which is substantially constant for
all speeds in the case of ball bearings, but increases
with the speed in an ill-defined manner in the case
of plain bearings. In addition there is the air
resistance, which increases with the speed. In
turbine work this can be shown to vary as the cube
of the speed, and the fifth power of the diameter.
In the present case its importance is somewhat
less (unless speeds are very high) and even at full
speed, when its maximum effect is obtained, the
total power required is apt to vary approximately
as the square of the speed for small variations.
Then the driving belt if present absorbs (according
to Goodman, " Mechanics," p. 349) 16 to 21 ft. lb.
per sq. ft. passing over each pulley, and probably
one may take about half this for guide pulleys.

The coefficient of friction is also of importance,
and is found to be variable. Fig. 38 is a chart
prepared by Messrs. Ferodo, Ltd., showing the

relation of μ to slipping speed in feet per minute, for various substances at a pressure of 20 lb. per sq. in. on a 0·7% carbon steel water-cooled drum. Cast iron heated up, aluminium rapidly tore. Ash appeared the best wood, and next to it came oak, elm and teak in that order. The coefficient

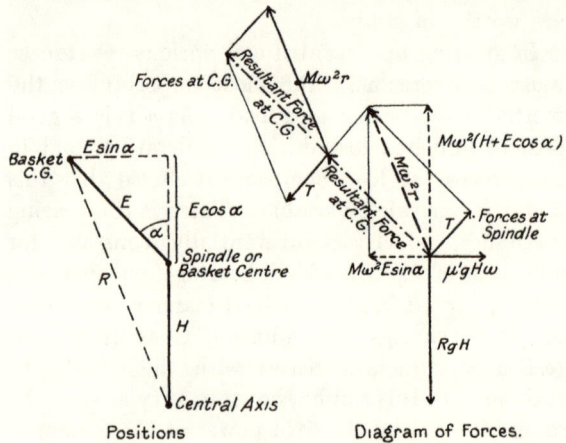

FIG. 38.—Forces on Self-balancing Centrifugal with Friction Present.

of friction of ash was 0·340 at 500 to 1,000 ft. per minute, 0·490 at 1,000 to 2,500 f.p.m.; 0·500 at 3,000 to 5,000 f.p.m., and 0·624 at 5,000–7,000 f.p.m. Elm charred very readily.

Having in view all the possibilities of variation it appears that any great refinement in calculations is of little value, and the final setting of the shoes is best made by trial, when their weight can be adjusted as required.

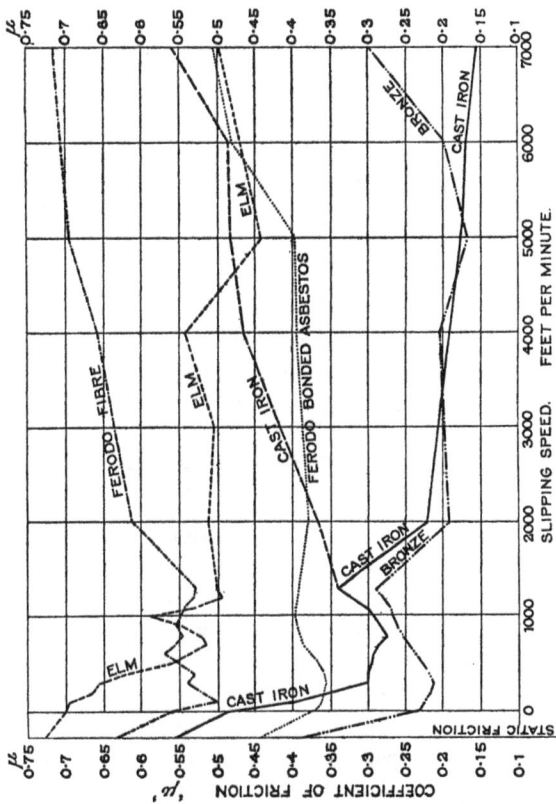

FIG. 39.—Alteration of Coefficient of Friction with Slipping Speed.

If a motor of low starting torque is fitted with a clutch, the shoes are often restrained by springs so that it can attain a fair speed before they engage. If we settle this speed we can reckon the pull of the springs, and obtain the net thrust of the shoes, and the torque exerted. A simple calculation gives the horse-power absorbed as :—

$$P_A = \frac{ML(N^2{}_k - N^2{}_0)\mu N_k D_c}{3 \cdot 08 \times 10^7} \qquad . \qquad (1)$$

Here P_A = horse-power absorbed.
$\quad$ M = total mass of shoes—lb.
$\quad$ N_k = constant speed of arms—r.p.m.
$\quad$ N_0 = speed at which shoes come into play.
$\quad$ L = length of arms (to c.g. of shoes), ft.
$\quad$ D_c = diameter of clutch face, ft.
$\quad$ μ = coefficient of friction.

It is usual to assume that the driving shaft comes almost immediately to full speed. This is not always the case, for if too heavy shoes are fitted, a motor of low starting torque may gain speed only slowly. If the motor has a high starting torque, current may be economised by designing the clutch to lock long before the motor reaches full speed. Space does not permit consideration of these cases or of the problems arising out of the reaction of the arm thrust on the shoes.*

Since the torque is equal for both parts of the clutch member, the horse-power delivered by the clutch is :—

$$P_A \cdot \frac{\omega}{\omega_k}.$$

* _Vide_ Patent Specifications 180,417, 206,699.

Where ω = speed of driven member. Radians per sec.

ω_k = constant speed of arms. Radians per sec.

Further, for the considerations given above, it is sufficiently accurate to take the power absorbed in friction as proportioned to the square of the speed, basing it on that taken at top speed.

The h.p. absorbed in friction is :—

$$P_F = P_M \frac{\omega^2}{\omega_k^2}. \qquad . \qquad . \qquad (2)$$

Where P_M = horse-power to maintain full speed.

The energy stored in the basket is :

$$\tfrac{1}{2} I_S \omega^2 \qquad . \qquad . \qquad . \qquad (3)$$

Where I_S is the moment of inertia (lb. ft.^{2}).

In a short time dt the speed increases by $d\omega$, and on equating the work done on the basket to the increase of energy we get, on neglecting squares of small quantities—

$$550g \left(P_A \frac{\omega}{\omega_k} - P_M \frac{\omega^2}{\omega_k^2} \right) dt = I_S \omega d\omega \qquad . \qquad (4)$$

Whence $\left(\text{remembering that}\right.$

$$\int \frac{dx}{a - bx} = -\frac{1}{b} \log (a - bx) \Big),$$

we get, for the time t_ω to reach any speed ω below ω_k,

$$t_\omega = \frac{I_S \omega_k^2}{550 g P_M} \log_e \left(\frac{P_A}{P_A - P_M \dfrac{\omega}{\omega_k}} \right) \text{secs.} \quad . \quad (5)$$

or $t_N = \dfrac{I_S N_k^2}{1 \cdot 61 \times 10^6 P_M} \log_e \left(\dfrac{P_A}{P_A - P_M \dfrac{N}{N_k}} \right)$ secs. (5a)

If the power to overcome friction were proportional to the speed, we should get

$$t_N = \frac{I_S N_k N}{1 \cdot 61 \times 10^6 (P_A - P_M)} \text{ secs.} \quad . \quad (5b)$$

This enables us to calculate the time taken to come up to speed and draw the acceleration curve.

Another interesting case is that of a centrifugal driven by a Pelton wheel. Here the power transmitted to the vanes is *

$$P_A \times \left(\frac{2f(V_J - V)V}{V_J^2} \right).$$

Here P_A = accelerating h.p. (water h.p.)

f = a factor depending on angle jet is turned through by buckets (modified by friction and spacing of buckets). In theory it equals 2 for a frictionless vane turning the jet through 180°.

V_J = velocity of jets. f.p.s.

V = Velocity of vanes = ωr.

ω = velocity of vanes. Radians per sec.

r = average radius of vanes. Feet.

We can equate work done to gain in energy as in the case of the clutch and obtain :—

$$550 \left\{ \frac{P_A 2f(V_J - V)V}{V_J^2} - P_M \frac{\omega^2}{\omega_M^2} \right\} g dt = I_S \omega d\omega \quad (6)$$

* Lea, "Hydraulics," p. 268. Jamieson, "Applied Mechanics," 4th Ed., Vol. II, p. 601.

Here ω_M = Full speed.   Radians per sec.
By integration we get

$$t_\omega = \left(\frac{I_S V_J^2 \omega_M^2}{550g(2P_A fr^2\omega_M^2 + P_M V_J^2)}\right) \log_e$$

$$\left(\frac{2P_A fr\omega_M^2 V_J}{2P_A fr\omega_M^2(V_J - r\omega) - P_M V_J^2 \omega}\right) \text{ in secs.}   (7)$$

We can simplify this by inserting the values Q_A, for the flow of water, in lb. per sec. during acceleration and Q_M for maintaining running against friction, etc., at full speed.

Then $$P_A = \frac{Q_A V_J^2}{2 \times 550 \times g}$$.   .   (8)

If V_J is equal for both jets, then

$$t_\omega = \left(\frac{2I_S \omega_M^2}{2Q_A fr^2\omega_M^2 + Q_M V_J^2}\right) \log_e$$

$$\left(\frac{2Q_A fr\omega_M^2 V_J}{2Q_A fr\omega^2_M(V_J - r\omega) - Q_M V^2_J \omega}\right) \text{in secs .}   (9)$$

It may be noted that the maximum efficiency of the Pelton wheel should occur a little below top speed, so that a slight decrease only of speed may bring additional torque into play to overcome any additional resistance which may occur.

Excessive power is usually required to complete acceleration under 2 minutes.   Three to 4 minutes may at times be allowed for large fixed spindle machines.   This rarely affects the total length of operation, and may be an advantage in the case of products which tend to become impermeable, since the pores remain open longer.

Applied Mechanics. Jamieson. Ed. 8, Vol. I. pp. 157, 314, 316, 323. Vol. IV. p. 135. Vol. V. pp. 444, 449–451, 456.

Dairy Chemistry. Richmond. (1920. Chap. xxvii.)

Engineering. " Centrifugal Machines—Power to Drive." (1903. Vol. LXXVI. p. 248.) " Centrifugal Machinery." Viola. (1903. Vol. LXXVI. p. 298.)
" Water-driven Centrifugals." Broadbent. (1921. Vol. III. p. 744.)

Dressing of Minerals, The. Louis. " Separation by Specific Gravity," chap. v.

Journal of the Institute of Brewing. " Centrifugal Separators and Filters for use in the Brewery." Heron. (1922. Vol. XIX. p. 498.)

Journal of the Oil and Colour Chemists' Association. " Supercentrifugal Force, and its Application to the Classification of Varnish and the Dehydration of Oils." Keable. (1922. Vol. V. p. 2.)

Chemical Age. " Centrifugal Extractors and Separators." Alliott. (1919. Vol. I. p. 717.)
" Centrifugal Extractors and Separators applied to the Chemical Industry." Broadbent. (1920. Vol. III. p. 652.)
" The Driving of Centrifuges." Broadbent. (1921. Vol. IV. p. 183.)

Journal of the Society of Chemical Industry. " Application of Centrifugal Force to Suspensions and Emulsions." Ayres. (1916. Vol. XXXV. p. 676.)
" Draining Crystals in a Centrifugal Machine." Drakeley and Martin. (1921. Vol. XL. p. 308т.)
" Centrifugal Draining—Efficiency in." Drakeley and Williams. (1922. Vol. XLI. p. 347т.)

Mathematic and Physical Papers. Stokes. (Vol. III. 1901. p. 59.)

Journal of the Textile Institute. " The Efficiency of a Centrifuge for Removing Surface Liquids from Cotton Hairs." Coward and Spencer. (1923. Vol. XIV. No. 1.)

Philosophical Magazine. " On the Motion of a Sphere in a Viscous Liquid." Allen. (1900, p. 324.) "Limitations imposed by slip and inertia terms on Stokes' Law." Arnold. (1911, p. 755.)

Power Laundry. " Hydroextractors and Allied Types of Machinery." Alliott. (1923. March and April, Nos. 485, 486, 487, 488.)

Proceedings of the Chemical Engineering Group (Society of Chemical Industry). " Centrifugal Dryers and Separators." Alliott. (1924. Vol. VI. p. 72.)

Spinning Tops. Crabtree.

Transactions of the American Institute of Chemical Engineers. " Centrifugal Force on Colloidal Solutions—The Effect of." Ayres. (1916. Vol. IX. p. 203.)

Transactions of the Institute of Chemical Engineers. Bibliography. (1923. Vol. I. p. 116.)
" Self-balancing Centrifugals." Alliott. (1924, Vol. II. p. 39.)

Vibration in Engineering. Frith and Buckingham.

Theory of Emulsions and Emulsification. Clayton. (pp. 136–193.)

LIST OF SYMBOLS

It has not been possible in every instance to avoid using the same symbol for different meanings; but such cases should be clear from the content and no confusion should arise.

Note.—For Greek Symbols see pp. 146 and 147.

Symbol.	Meaning.		1st Reference.
a	Radius of particle.	Cms.	80
a_c	Critical radius of suspended particle.	,,	81
B	Torque modulus of buffer.	Ft. lbs. per radian deflection.	127
B'	Net buffer torque when basket and load are removed.	,,	128
B_s	Force on the spindle, due to unbalanced load.	Lb.	126
C	Centrifugal effect at any diameter D ft. *i.e.* ratio to gravity.	Lb. per lb. or Gms. per gm.	12
C_0	Centrifugal effect corresponding to diameter D_0 and to K_0.	,, ,,	85
C_P	Centrifugal effect at periphery.	,, ,,	11
c.g.	Centre of gravity.		
d	An indefinitely small increment.		
D	Any diameter.	Ft.	12
D_C	Diameter of the clutch face.	,,	138
D_H	Diameter of the heavy liquid, overflow.	Any Units.	89
D_L	Diameter of the light liquid, overflow.	,, ,,	89
D_L	Diameter of the inner face of load.	Ft.	12
D_O	,, ,, reference diameter in centrifuge. corresponding to K_0 and C_0.	Cms.	85
D_P	Diameter of basket periphery.	Ft.	11
D_S	,, of separating plate.	Any Units.	89
e	Base of Napierian logarithms (2·7183)		84
E	Eccentricity of basket c.g.	Ft.	125
f	% of fat in skim milk.		90
f	Factor depending on angle through which jet is turned, and on vane friction, etc.		140
f_i	% of initial fat in milk.		92
F	% of fat in cream.		91
F_s	Viscous resistance of liquid.	Dynes.	80
g	Gravity constant. 32·2 or 981.		x
h	Height of liquid surface above the vertex.	Ft. or Cms.	79
h	Height of top layer of suspended particles.	Any Units.	82
h_0	Height to which particles sink after an indefinitely long period.	,, ,,	82
h	Height corresponding to number of particles equal to K_1 per cubic cm.	Cms.	84
h	A variable depth.	,,	84
h'	A fixed total depth.	,,	84
HP.	Horse-power.		92
H	Deflection at c.g.	Ft.	125
H_D	Diameter of holes in periphery.	Inches.	17
H_P	Pitch of holes in periphery.	,,	17
I_B	Moment of inertia of basket about the centre of buffer.	Lb. ft.².	127

144

Symbol.	Meaning.		1st Reference.
Is	Moment of inertia of basket about the spindle.	Lb. ft.².	127
I'ʙ	Moment of inertia of balanced portions only of revolving system about buffer centre.	,, ,,	129
I's	Moment of inertia of balanced portions only of revolving system about the spindle.	,, ,,	129
k	Factor (3 × 10¹³ approximately).		84
k₁	Separator constant.		91
k₂	,, ,,		91
k₃	,, ,,		91
kf	Factor (usually 0·5).		80
K	Constant.		29
K₀, K₀·₅, K₁	Volume concentrations at 0, 0·5, and 1 cm. from datum (volume percentage).		85
K	No. of particles per cubic cm. at height h cms. above datum (or volume percentage).		84
			84
Kᴅ	Radius of gyration of basket and load about a diameter through c.g.	Ft.	123
Kₜ	Initial concentration of suspended particles. I.e. no. of particles per cubic cm. or volume percentage (see h). In equation (15) it is to be taken as a weight percentage.		84
Kₒ	No. of particles per cubic cm. at datum (or volume percentage).		84
Ks	Radius of gyration of basket and load about the spindle.	,,	123
l	Length corresponding to force P.	,,	128
L	,, of suspension rods.	,,	126
L	,, spindle-buffer centre to c.g.	,,	123
L	,, of clutch arms—axis to c.g. of shoes.	,,	138
m	Mass of suspended particle.	Grams.	84
M	,, of basket and load.	Lb.	123
M	Total mass of shoes.	,,	138
M₁	Mass of load per foot of basket depth.	,,	15
Mᴘ	Mass of outer suspended pan.	,,	126
N	Revolutions per minute.		11
Nᴀ	Avogadro's number (62 × 10²²).		84
Nᴄ	Critical speed of basket.	R.P M.	125
Nʜ	Speed of basket for deflection H.	,,	125
N-ʜ	,, ,, ,, ,, -H.	,,	125
Nᴍᴄ	Critical speed giving minimum critical period.	,,	131
Nₒ	Speed at which shoes come into play.	,,	138
Nₖ	Constant speed of arms.	,,	138
P	Force to deflect basket 1 inch at distance l feet from buffer centre.	Lb.	128
Pᴀ	Accelerating water horse-power.		140
Pᴀ	Horsepower absorbed by clutch.		138
Pꜰ	,, absorbed by friction.		139
Pᴍ	,, to maintain full speed.		139
Pʟ	Pressure exerted by liquid load on rim.	Lb. per inch.².	15

Symbol.	Meaning.		1st Reference.
Ps	Pressure exerted by solid load on rim.	Lb. per inch.^{2}	15
Pp	Pressure on periphery due to own mass.	,, ,,	17
Q	Rate of flow.	Gallons per hr.	91
QA	,, ,, of water during acceleration.	Lb. per sec.	141
QM	Rate of flow of water to maintain full speed.	,, ,,	141
r	Radius of liquid surface at height h above vertex.	Ft. or Cms.	79
r	Radius at which out of balance weight U is situated.	Ft.	129
r	Average radius of vanes.	,,	140
R	Amount of remaining liquor per 100 lb. of dry substance.	Lb.	28
R	Gas constant (83·2 × 10^6).		84
R	% of cream recovered.		92
R	Restoring force of buffer.	Lb. per ft. of deflection at c.g.	125
R.P.M.	Revolutions per minute.		
SL	Density of Load.	Lb. per Ft.^{3}	12
SP	Average weight of the periphery.	Lb. per inch.^{3}	17
t	Time of separation.	Seconds.	86
t	Temperature of milk.	$^{\circ}$C.	90
tN	,, to reach speed N r.p.m., up to top speed. Nk.	,,	140
t_ω ·	Time to reach any speed ω up to ωk.	,,	140
T	Absolute temperature.	Deg. Cent.	84
T,T	Forces at ends of eccentric couple arm E.	Lbs.	132
Tp	Tensile stress in periphery.	Lb. per inch.^{2}	17
U	Out of balance weight—at periphery or at radius r.	Lb.	129
v	Velocity of vanes.	Ft. per sec.	140
v_c	Critical velocity of suspended particles.	Cms. per sec.	81
v_s	Velocity of subsidence.	,,	80
VJ	,, of jets.	Ft. per sec.	140
V$_O$	Speed at point of discharge.	,,	92
VP	Linear, surface, or peripheral speed.	,,	12
Z	Thickness of periphery.	Inches.	17
α	Angle of inclination of spindle.	Radians.	127
α	An omnibus symbol.		84
β	,, ,,		85
$\theta(\beta), \phi(\beta)$	Functions of β.		86
η	Viscosity.	Dynes per cm.^{2}.	81
ω	Angular velocity of basket.	Radians per sec.	11
ω	Speed of driven member of clutch.	,, ,,	139
ω	Velocity of vanes.	,, ,,	141
ωk	Constant speed of clutch arms.	,, ,,	139
ωM	Full speed of Pelton vanes.	,, ,,	141
Ω	Precession.	,, ,,	127
Ω_O	Direct precession.	,, ,,	121
Ω_R	Reverse precession (i.e. opposite to rotation).	,, ,,	121
μ	The micron (0·001 mm., $\mu\mu$ 0·000001 mm.).		87
μ	Coefficient of friction of clutch.		136

Symbol.	Meaning.		1st Reference.
μ'	Kinematic viscosity.	Cm.^{2} per sec.	80
μ'	Frictional drag.	Lb. per ft. per sec.	132
π	Constant (3·1416).		
ρ	Density of liquor (or specific gravity).	Gms. per cubic cm.	80
ρH	,, heavy liquor.	,, ,,	89
ρL	,, light liquor.	,, ,,	89
σ	,, suspended particles.	,, ,,	80
		,, ,,	80

INDEX

6

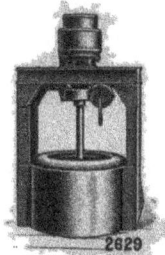

PALM OIL FACTORY

ERECTED BY

CULLEY EXPRESSORS LTD.

14, ST. MARY AXE, LONDON

**PALM OIL, FREE FROM WATER AND
FOOTS, FROM STERILISED FRUITS
IN LESS THAN TWENTY MINUTES**

COMPLETE INSTALLATIONS
: SUPPLIED AND ERECTED :

Sharples

Super-Centrifugal

Dehydration

OF

BONE GREASE

BUTTER FAT

COAL TAR OIL

CRUDE PETROLEUM

CYLINDER STOCK

REFINED LINSEED OIL

ETC.

GIVES BEST RESULTS

Catalogue from

SUPER-CENTRIFUGAL ENGINEERS, Ltd

Aldwych House

London, W.C. 2

TELEPHONE :   TELEGRAMS :
HOLBORN 3111 (8 LINES).   "SUPERSPIN, ESTRAND, LONDON."

MANLOVE ALLIOTT & CO LTD

FILTER PRESSES

NOTTINGHAM

www.ingramcontent.com/pod-product-compliance
Lightning Source LLC
Chambersburg PA
CBHW021425180326
41458CB00001B/136